THE
EVERYTHING®
GUIDE TO
MOBILE APPS

Dear Reader,

It's both exciting and terrifying to think about the mobile technology opportunities that lay ahead. It's no coincidence that companies everywhere are scrambling to produce smartphone and tablet apps aimed at getting their content, services, and brands in the hands of customers.

It's time for us all to take out and dust off Henry Chesbrough's milestone book, *Open Innovation: The New Imperative for Creating and Profiting from Technology*, and read it with mobile in mind. Chesbrough famously argued in 2003 that companies must be open to good ideas wherever they find them, and should "leverage multiple paths to market—even if the path to success is through another company's business."

We're sure Chesbrough would agree that this approach also powers the new App Economy, where companies and developers alike are creating value-adding mobile apps. Alone or together with partners, you have to plan for success.

But app developers are not just expected to master all the business and technology challenges of a multichannel, multidevice, multimedia world. In many cases, they must take on new responsibilities on behalf of their clients and partners, provide the ideas, build the apps, and drive the social media and marketing that will benefit everyone in their business ecosystem.

It's a mammoth task, which is why we have brought together the insights of twenty-five-plus mobile technology professionals, practitioners, and pundits to identify market trends, best practices, and key lessons learned in developing and distributing mobile apps. Our purpose is to provide you clear direction and critical information, equipping you to profit from your app venture. Think of this guide as a starting point to a detailed roadmap, one that will allow you to plot the transformational path your app business needs to succeed in this exciting new App Economy.

Strap yourself in. It's going to be a terrific ride!

Peggy Anne Salz and Jennifer W.

Welcome to the EVERYTHING® Series!

These handy, accessible books give you all you need to tackle a difficult project, gain a new hobby, comprehend a fascinating topic, prepare for an exam, or even brush up on something you learned back in school but have since forgotten.

You can choose to read an Everything® book from cover to cover or just pick out the information you want from our four useful boxes: e-questions, e-facts, e-alerts, and e-ssentials.

We give you everything you need to know on the subject, but throw in a lot of fun stuff along the way, too.

We now have more than 400 Everything® books in print, spanning such wide-ranging categories as weddings, pregnancy, cooking, music instruction, foreign language, crafts, pets, New Age, and so much more. When you're done reading them all, you can finally say you know Everything®!

QUESTION

Answers to
common questions

FACT

Important snippets
of information

ALERT

Urgent
warnings

ESSENTIAL

Quick
handy tips

PUBLISHER Karen Cooper

MANAGING EDITOR, EVERYTHING® SERIES Lisa Laing

COPY CHIEF Casey Ebert

ASSOCIATE PRODUCTION EDITOR Mary Beth Dolan

ACQUISITIONS EDITOR Lisa Laing

ASSOCIATE DEVELOPMENT EDITOR Eileen Mullan

EVERYTHING® SERIES COVER DESIGNER Erin Alexander

Visit the entire Everything® series at *www.everything.com*

THE
EVERYTHING®
GUIDE TO
MOBILE APPS

A practical guide to affordable mobile app development
for your business

Peggy Anne Salz and Jennifer Moranz

A adamsmedia
Avon, Massachusetts

This book is dedicated to my husband, Peter Salz, without whose encouragement, support, and sense of humor, none of this would have been possible, and to my mother, Ruth Trautman, who started the flame. —Peggy

This book is dedicated to my LOML, Jonathan Shambroom—your love, laughter, and monkey sounds made this possible—and to my mother, Judy DeVoto, for your never-ending support of my passion for writing. —Jennifer

An Everything® Series Book.
Everything® and everything.com® are registered trademarks of F+W Media, Inc.

Published by Adams Media, a division of F+W Media, Inc.
57 Littlefield Street, Avon, MA 02322. U.S.A.
www.adamsmedia.com

ISBN 10: 1-4405-5533-8
ISBN 13: 978-1-4405-5533-6
eISBN 10: 1-4405-5534-6
eISBN 13: 978-1-4405-5534-3

Printed in the United States of America.

10 9 8 7 6 5 4 3 2 1

This publication is designed to provide accurate and authoritative information with regard to the subject matter covered. It is sold with the understanding that the publisher is not engaged in rendering legal, accounting, or other professional advice. If legal advice or other expert assistance is required, the services of a competent professional person should be sought.
—From a *Declaration of Principles* jointly adopted by a Committee of the American Bar Association and a Committee of Publishers and Associations

Many of the designations used by manufacturers and sellers to distinguish their products are claimed as trademarks. Where those designations appear in this book and F+W Media was aware of a trademark claim, the designations have been printed with initial capital letters.

*This book is available at quantity discounts for bulk purchases.
For information, please call 1-800-289-0963.*

Contents

Top 10 Things You Need to Know about Mobile Apps **10**

Introduction **11**

01 The App Defined / 13

What Is an App? **14**

Types of Apps **15**

How Do Apps Work? **18**

Today's Smartphones **20**

Where to Start? **22**

Building a Mobile Strategy **23**

Make Them an Offer **24**

The App Environment **26**

02 App Market by the Numbers / 29

The Numbers Game **30**

App Interaction on the Rise **33**

Hot Consumer Trends **35**

Staying On Top **36**

Will Your App Stack Up? **37**

03 Know Your Audience / 41

Segmenting the Market **42**

Mobile Demographics (Ages 8–34) **42**

Gen-X: Ages 35–54 **46**

Digital Moms **46**

Boomers Are the Boom Market **48**

You Are Mobile **53**

04 You Are Not Alone / 55

Resources for Building **56**

Strength in Numbers **58**

The App Economy Is Here! **60**

Reality Bytes **65**

Developer Obstacles **66**

05 What You Need to Know Before Building / 69

Value, Benefit, and Cost **70**

Design Costs **71**

Development Costs **72**

Risky Business **73**

What Are the Payment Terms? **77**

Owning What Matters Most **79**

Adjusting to Change **82**

06 Developing Your App / 83

Who Will Make Your App? **84**

Should You Go In-House? **84**

The Cost of Development Methods and Models **86**

Factors That Impact Cost **89**

How to Find a Good Developer **91**

Clear Communications **94**

07 Designing Your App / 95

Starting Your Development Career **96**

Be Consistent **97**

Consider Context and Usage **98**

Multiple Platforms and Devices **100**

Catering to Global Capabilities **101**

Visualize Your App **104**

User Experience **105**

08 Personal Privacy Is Paramount / 107

Protecting the Consumer **108**

Privacy Is a Pivotal Issue **109**

Gaining Trust Through App Best Practices **111**

The Privacy Imbalance **115**

09 Testing, Testing / 119

The Need for Testing **120**

Test Everything Everywhere **121**

Be Ready to Change Perspectives **122**

What to Watch For **124**

Crowdsourcing for Constructive Feedback **125**

Testing for Local Flavor and Audiences **127**

Think First **128**

10 Shop Till You Drop / 129

The Evolution of Mobile Shopping **130**

Understand Your Mobile Shopper **130**

Consumer Motivation **132**

Location Is Key **135**

Leveraging Location **138**

Monetizing Location **141**

11 App Stores and More / 145

Choosing Your App Store(s) **146**

Submitting to the Major App Stores **149**

Tackling the App Approval Process **151**

Getting the Go-Ahead, or Not **152**

Understand and Harness Momentum **153**

App Store Shortcomings **155**

A Store for Enterprise Apps **158**

Set Your Priorities **159**

12 App Monetization Models / 161

Monetization Models Explained **162**

App Stickiness Matters **166**

How to Make Money from Your Apps **167**

The Price Is Right **168**

Business Models for Game Apps **169**

Choosing the Right Model for Your Game **170**

13 Mobile Marketing and Advertising / **173**

Mobile Marketing Defined **174**

The Power of Mobile Marketing **174**

A Word about Location-Based Marketing **178**

Mobile Advertising 101: The Basic Lingo **179**

Integrating Ads into Your App **181**

Advertising Challenges **183**

Maximizing Your In-App Advertising Revenue **184**

Reaping the Benefits **185**

14 Getting Discovered Across All Channels / **189**

It's Retail 101 All Over Again **190**

The Discovery Dilemma **192**

Get Your Industry Street Cred **195**

Promoting Your App **198**

Creating Hype Through Reviews **200**

Rethinking App Discovery **202**

15 Achieve Lasting Impact and Loyalty / **205**

Encouraging Lasting App Engagement **206**

The Power of Push **206**

Extend the Life of Your App **209**

Good Push Drives Great Results **210**

Serve Your Loyal App Users **213**

16 Get Seriously Social / **219**

Build Your Community **220**

Social Media Crash Course **224**

Social Media Skills Test **227**

Networking **228**

Time Is on Your Side **230**

17 App Maintenance / **237**

Postlaunch Challenges **238**

Map Out Your Maintenance **239**

Crash Reporting Defined **240**

Why Crash Reporting Is Important **242**

Reasons Why Apps Crash **243**

Platform Problems **246**

18 Success Breeds Success / **249**

The Rise of the *Appreneur* **250**

Project NOAH: A Case Study **251**

LevelUp: A Case Study **254**

Gigwalk: A Case Study **257**

From Tweets to Living Photos **260**

New Rules, New Game **262**

19 **The Future of Apps / 263**

Looking Forward **264**

Apps Power Machines **264**

Apps Improve Customer Service **266**

Artificial Intelligence Gets Simpler **269**

Apps for Good **274**

Appendix A: Glossary of Mobile Terms **277**

Appendix B: Contributor Biographies **284**

Index **297**

Acknowledgments

A special shout-out to all our guest contributors and mobile pundits, with a hat tip to Tomi Ahonen for excellent mobile stats; Ken Herron for his invaluable insights around social media; Eileen Mullan for being a good listener and a great editor; and Jez and Debi Harper at Tús Nua Designs for keeping it real, sharing their passion, and giving their all to make sure this book sets the bar. It has been such a pleasure and an experience working with you all and the result is a resource we can all be proud of!

Top 10 Things You Need to Know about Mobile Apps

1. An app may be overkill. In some cases, a mobile-optimized website will do the job just fine.

2. You can't go it alone! It's only by cultivating partnerships and alliances that you can effectively create a network of support to help market your app.

3. Make your device platform choice carefully.

4. Build relationships, not just apps.

5. App marketing must be a constant focus, so keep your eye on the prize—always.

6. Know your audience. Mobile devices are fiercely personal and people expect customized experiences.

7. Personal privacy is paramount. You must tell your users up front what kind of data about them is being collected.

8. Be aware of billing solutions. You should choose your payment methods and mechanisms wisely.

9. Postlaunch problems abound, and a multitude of variables can affect how your app performs.

10. Think big and be useful. Apps serve people, so expect the next wave of innovation to focus on utility, not novelty.

Introduction

LOOK AROUND AND IT'S clear that conditions are coming together for a "perfect storm" that will rip across the mobile app space, paving the way for mobile apps that will enable everything—from games and entertainment to education and health care to retail and daily productivity—and leave an indelible mark on the daily lives of many.

It all links back to the power of mobile, a true mass-medium that has been growing for years. Think of the printed word from the 1500s, sound recording from the 1900s, cinema from the 1910s, radio from the 1920s, TV from the 1950s, and Internet from the 1990s. The power of mobile lies in its simplicity, ubiquity, and incredible reach. Approximately three-quarters of the world's population now has access to a mobile phone. That's higher than the number of people who have clean drinking water (!). In a little over a decade, the number of mobile phone users has increased six times, from 1 billion in the year 2000 to a whopping 6 billion in 2012.

From the first mobile music ringtone, developed by Finnish technologist and entrepreneur Vesa-Matti "Vesku" Paananen in 1998, the early evolutionary path of mobile technology has been sharply focused on encouraging the passive consumption of prepackaged content. Not surprisingly, mobile downloadable content—from ringtones to wallpapers—became a multibillion-dollar industry.

At this point, mobile is about much more than buying novelty content. The advance of mobile apps now *empowers* people to call the shots. Armed with apps they have the means to make decisions, create and customize mobile experiences, conduct commerce and business, and bridge the digital and physical worlds around them. From life-simplifying tools to life-saving medical advice, mobile apps are poised to change the nature of commerce, banking, education, health care, news reporting, and political participation.

A study released by the Application Developers Alliance declares there is "no end in sight for app market growth." Indeed, global demand for apps continues to exceed analyst expectations, with nearly half of the population in the United States alone having downloaded an app. The supply side of the equation also breaks its share of records as innovations in devices (the launch of the iPad mini), platforms (the long-awaited arrival of Research In Motion's BlackBerry 10 operating system), use cases (branded apps for marketing and retail, and life-simplifying apps for health and wellness) drive a new phase of growth and creativity.

And if you think you have to be a developer to get in on the action, think again. The app ecosystem is also quickly evolving to accommodate people who are not just programmers. The full-scale arrival of app component marketplaces, the growing availability of easy-to-use programming tools, and the advance of cloud-based support and distribution are making it possible for anyone, anywhere to build an app.

From job creation, driven by companies that must scale up to meet massive demand, to an avalanche of opportunity unleashed as enterprises contract apps to boost productivity, support marketing, and improve customer service, apps are emerging as the growth engine of a new and vibrant economy. Connect the dots, and you'll see that the App Economy has officially arrived! Now it's time for you to get to work and reap the profits.

Turning your app into a serious business (if you're an app developer) or architecting an app to achieve sustained market presence from your product or service (if you're a business) requires you to know your audience, understand your market, and plan for success. In a way, it's Retail 101 all over again, but there's a catch: You have to do more than sell them; you have to convince users to insert your app into their daily routines. After all, app downloads mean nothing if people don't use them again and again.

Because mobile devices are fiercely personal, they present an ideal means to reach customers during every step of their daily journey to deliver value, encourage interaction, deepen engagement, boost customer loyalty, and—ultimately—recruit true brand advocates and app fans. At the other end of the spectrum, apps help us get things done. They inform, entertain, communicate, advise, manage, streamline, simplify, and execute. If you want to create a winning app, you must have a broad base of knowledge that allows you to identify your target audience and delight them again and again.

CHAPTER 1

The App Defined

On April 3, 1973, a Motorola researcher and executive, Martin Cooper, changed history. On that day, he made the first successful telephone call from a handheld device to Dr. Joel S. Engel at AT&T Bell Labs. Since then, the world has seen an explosive growth in the mobile phone market.

The mobile phone has evolved from the kitschy box phones of the 1980s, to limited-capability feature phones, and now to the beloved smartphones of today. As the capabilities of mobile phones have developed over the years, a new way to consume and promote content has evolved as well: the app!

The term *app*, which was traditionally used in the PC market, has been adopted throughout the consumer market to refer to the method used to communicate, entertain, educate, shop, and spend money from your mobile phone. The mobile app ecosystem is quickly becoming a numbers game as every facet of the marketplace continues to see unparalleled growth. There are an astounding number of apps for consumers to download in today's app market, and there are so many more opportunities to make it big in mobile on the horizon.

What Is an App?

A mobile app is a piece of software specifically designed to run on a mobile device, such as a smartphone or tablet. The app is usually downloaded and installed by the device owner, and once installed, a mobile app typically operates in tandem with the device's native, or installed, operating system (OS). Very often, this allows the app to take advantage of some of the mobile device's features.

If you are having trouble imagining this whole scenario, it helps to think of a smartphone or tablet as a portable computer. For example, on your computer at home, you download a picture from a family member's e-mail onto your hard drive. You are then able to view the picture from your computer without having to connect to your e-mail. The picture lives on the hard drive, just like when you download an app onto your smartphone, the app lives on the smartphone and you can access it whenever you want.

App Basics

There are a number of confusing terms linked to mobile apps that you will often hear: *native*, *hybrid*, *wrapper*, and *thin client*. Unless you know exactly what you are looking to achieve when building a mobile app, these terms are relatively useless.

At this moment, all you need to know is that native, hybrid, wrapper, and thin client mobile apps are basically the same thing. They are based on a small application (small in data size, so only a little bit of information) that is downloaded and installed on a mobile device. Content such as pictures or video is then pulled over the Internet via a mobile data connection (Wi-Fi), and once the content is embedded in the device (your smartphone), the data connection can be closed and the content viewed offline (when you aren't connected to the Internet).

In essence, you are downloading a piece of content to your phone. The option to view content offline is attractive to some app publishers, as it offers their customers a way to access their mobile app when a data connection isn't available. For example, someone who takes an underground subway to work will not be able to receive a wireless Internet connection during their commute, so unless the content they want to view is already on their phone, such as existing within a downloaded app, they are in for a long, boring ride.

QUESTION

What should an app offer?
An experience that is more than a customer can get from an advanced mobile website. Having only one reason for placing an icon on your customer's mobile device home screen is not good enough. This can be achieved very easily with a mobile website.

Once developed, mobile apps are typically distributed via a third party, commonly known as an app store or marketplace. Unfortunately, you can't just post your app to the market without any hurdles. For almost all apps, there is a review and approval process, which adds time and complexity to getting any service to market. For some stores there is even a subscription fee before you can submit an app for review.

Types of Apps

If your ambition is to gain a presence on the home screen of a person's device in order to encourage usage and loyalty, you do not necessarily need a mobile app. Gaining a presence for your brand and service on many mobile device home screens can be achieved very easily using a mobile web approach.

Mobile Web Apps

Mobile web refers to content that is viewed through a smartphone's web browser. If you have a mobile version of your website, any mobile device owner with an Internet browser installed (Internet Explorer, Mozilla Firefox) can view your mobile website.

If your goal is to reach your customers and engage with them on a daily basis, an app might be the best choice for you. If you plan on asking your audience to access your mobile app multiple times a day, it is important to provide them with an easy-to-access method of engaging with your branded mobile experience. This is why building a mobile app is the appropriate choice.

There are many factors that will determine your approach to mobile technology. Mobile web is not better than mobile apps, or vice versa—no approach is better than the other. It really depends on: what you are looking to achieve, how much money you have to invest, how often you need to add new features and update your service, how loyal your typical customers are, how you want to win new customers. These will all factor in defining the right approach for you.

Mobile apps offer advantages such as their ability to use features of the mobile device and to store larger levels of content for viewing offline. Another advantage is they are often faster than the mobile web so you can find what you want more quickly. The disadvantages are that they need to be developed for individual devices and operating systems, they are more expensive to develop and maintain, distribution is often linked to a third party, and general discovery through search is far more restricted, as content can't easily be presented to search engines.

A mobile app can be used to promote loyalty among users. This is only relevant if customers come back to you and your brand on a regular basis. For many brands from retail through to restaurateurs, that is simply not the case. If a customer typically has a simple interaction with your brand— buys a product from time to time or wants to find out where your store is located—an app can sometimes become a barrier. The need to download can deter a customer very easily.

Native Apps

Native apps look great and can be accessed without a web connection. Depending on your audience, you will need to develop and maintain an app for each platform and version you are looking to target. Building and maintaining multiple apps can be a drain on resources and expenses.

Native apps are distributed through a range of app stores either platform-specific or third-party app stores. Most app stores mandate a 30 percent revenue share and the use of their own billing system. One of the major challenges presented with app stores is discovery, so you need to understand how you can make your app visible among the other millions of apps out there. For example, you can run your own marketing campaigns to push users to your app in the app store and use different techniques to boost your rankings in the app store.

HTML5 Apps

Building your apps using HTML5 gives you ultimate control and flexibility. You can market directly to your consumers without the need to fight for position in app stores and have control over the payment solution you integrate.

HTML5 apps can be written once and deployed across all of the latest smartphones, with only a small amount of adaptation needed for each of the individual platforms, as well as PCs and Smart TVs. Development and wide deployment is therefore fast and cost effective, unlike native apps, where a new version is required for each platform. They can also be added as an icon onto the desktop just like a native app.

HTML5 is a new, fast-moving set of standards, so there are still some things you may need native code for, and due to the nature of how HTML5 is implemented, many observe that it is unlikely to ever be a replacement for native apps, as it will never be able to offer the same levels of performance or functionality as those offered through native app development. HTML5 works best on the latest smartphones.

Remember that if you choose your own method of monetization you will not be able to package up your HTML5 app and distribute it through Apple iTunes or other popular app stores, as they enforce the use of their own payment technology and take their 30 percent share.

Hybrid Apps

A hybrid app is basically an app developed in combination with HTML5 and native technology. While many confuse a hybrid app with a native app, there is a fundamental distinction: A hybrid application is built using web technology, and then wrapped in a platform-specific shell. The native shell makes the app look like native apps and, more importantly, makes it eligible to enter the app stores. In addition, app developers can also build in some of the native functionalities into a hybrid app, allowing it to access some of the native APIs (Application Programming Interface) and use device-specific hardware features, such as location and built-in sensors.

Cross-Platform Apps

Developers that want to reach all people on all platforms can choose to make a cross-platform app. Cross-platform frameworks—which include Appcelerator's Titanium, Rhodes, and PhoneGap—are designed to limit the work that a developer or development team has to put into creating apps for iOS, Android, BlackBerry, Windows Phone, and beyond.

As with any development strategy, there are pros and cons to taking a cross-platform approach to mobile application design and development. The idea behind most cross-platform frameworks is to limit development time by having users write their code in one language that can easily be used by multiple platforms. This is where a cross-platform framework delivers significant benefits. Rather than having to write the specific action or sequence for each platform, a developer can just write the code once and then reuse those bits in later projects or on other platforms. But there is a downside to consider. The cross-platform tools can be limiting and lock you into a vendor relationship where, if you want to switch to another platform, the code you wrote before is likely not going to be reusable without a lot of work.

How Do Apps Work?

As previously noted, a mobile app is a small piece of software installed in a user's mobile device that performs a specific function. During the download process, or potentially during the first use of the mobile app, a user will be prompted to accept terms and conditions. These are required to obtain consent to perform some of the tasks that the app may support. Core areas that are typically covered are: privacy, the ability to capture and use personal information, consent to use device location, and payment or commerce.

Once installed, the mobile app is typically presented as an icon on your customer's mobile device home screen. When they click on the icon, the application typically takes a few seconds to launch. This start time will enable the application to use the built-in features of the mobile device. For instance, a navigation service will automatically identify your customer's current location and reveal their position on a map. This is a case of the mobile app using the device's global positioning or GPS capability. A publisher can use the device's data connectivity to push content to the device so the latest content is available to customers.

Updating Content

The mobile device will search for or request content through the data connection and will fetch information over a web or Internet connection. One core advantage to a mobile app is that a high level of content and information can be stored or cached for viewing offline. Although a certain level of data can be cached in a mobile web environment, this amount is certainly not comparable to the capabilities of a mobile app.

Updates typically will require a customer to download a new version to replace the old mobile app. On the publisher's side of things, new versions will need to be submitted to the app store or marketplace for approval. This is a good way to add new features but can bring issues if not carefully thought through. If not managed properly, you might update your app so frequently that your customers will have download atrophy, or even worse, you will have an app in the app store with poor functionality or bugs and your customers will have a terrible experience. Customer experience must be managed, information must be stored, and any service personalization must be protected.

ESSENTIAL

One of the best ways to get your app noticed is to describe the app and the value it delivers. Get this right and people will be encouraged to download your mobile app to their mobile device.

Today's Smartphones

For many, today's smartphones seem more like mini-computers than the traditional dial-and-talk phones. That's because, in essence, they are! As mobile technology continues to advance, anything someone would normally be able to do only on their personal computer can now be done everywhere they go.

How Smartphones Work

Smartphones are built on a mobile computing platform and combine the functionality of a phone, portable media player, digital camera, video camera, and a GPS unit. The phone itself contains touch screens, Internet browsers, high-speed data access via Wi-Fi, and mobile broadband.

Smartphones contain an operating system (OS), similar to a personal computer (PC). The most common OSs known today are Google's Android OS, Apple's iOS, RIM's BlackBerry OS, Samsung's Bada OS, Microsoft's Windows Phone OS, HP's webOS, and the Linux MeeGo. The OS is what allows the phone's user to do everything they need to do, from taking pictures to making calls, checking e-mail, and watching the latest YouTube video. These operating systems are key to determining where a business should focus their efforts when considering building a mobile app. They are also essential factors in deciding where to focus your development time, energy, and resources.

FACT

The first Apple iPhone was unveiled by Steve Jobs, who was then the CEO of Apple, on January 9, 2007. A few months later in June 2007, the iPhone was released to the public, and the rest is history. Now smartphone growth is taking off!

According to Mary Meeker's *2012 Internet Trends Report* from Kleiner Perkins Caulfield Byers (*www.kpcb.com*), at the end of 2011, there were 953 million smartphone subscriptions. According to an IHS iSuppli Wireless Communications Market Tracker Report (*www.isuppli.com*), smartphones are expected to account for the majority of global cell phone shipments in

2013, two years earlier than previously predicted. Smartphone shipments in 2013 are forecast to account for 54 percent of the total cell phone market, up from 46 percent in 2012 and 35 percent in 2011. The year 2013 will mark the first time that smartphones will make up more than half of all cell phone shipments. By 2016, smartphones will represent 67.4 percent of the total cell phone market.

In addition to this growth in smartphones through 2016, it is important to understand the breakdown across the growth of the different OSs so that you are able to make informed decisions about where to focus your mobile app development efforts.

According to the *Worldwide Quarterly Mobile Phone Tracker Report*, published by International Data Corporation (IDC, *www.idc.com*), Android and iOS will see the bulk of the growth. IDC found that:

- Android will remain the most shipped smartphone operating system over the course of the five-year forecast.
- iOS will continue its run thanks to the momentum for the iPhone 4S in North America, Western Europe, and Asia/Pacific, specifically China. Growth will moderate over the five-year forecast given the large base Apple has accumulated.
- Windows Phone 8 will gain share despite a slow start. IDC expects Windows 8 to be the number two operating software with more than a 19 percent share in 2016, assuming Nokia's growth in emerging markets is maintained.
- There will continue to be a market for BlackBerry OS–powered devices. This is true in emerging markets, for example, where users are looking for affordable messaging devices. However, the gap between the BlackBerry OS and its primary competition will likely widen over the forecast as the mobile phone market becomes increasingly software/app-oriented.

In regard to Nokia, the end of Symbian (Nokia's OS) as a widely used smartphone OS came in 2011 when Nokia said all of its smartphones would eventually be powered by Windows Phone OS. This announcement precipitated an Osborne-like effect that resulted in a sharp decline in Symbian's market share. It also led to share gains for competitive operating systems, namely Android and iOS. IDC expects Symbian-powered smartphone shipments to all but cease by 2014. Clearly, Nokia and Microsoft need to quickly switch Symbian OS user allegiances to Windows Phone 7 in order to maintain relevancy in the smartphone race.

Application Programming Interfaces (APIs)

One unique aspect of smartphone OSs is their ability to leverage advanced application programming interfaces (APIs). An API is a specification that is used as an interface by software components to communicate with each other. An API may include specifications for routines, data structures, object classes, and variables. APIs enable the phone to run third-party applications, or what are now commonly referred to as "apps."

For example, if you own an Android smartphone and want to run an application you downloaded from an outside source (perhaps a popular game), the API acts as the middleman between your phone's OS and that application. Without the API to negotiate between your phone's OS and the game, you would not be able to play. This process is what makes the consumer experience and new opportunities for developers extremely interesting.

ALERT

Make sure your app works on new devices and updated operating systems. There is nothing more frustrating for a customer than to suddenly find the mobile app that they have been using does not work on the new device that they have just bought.

Where to Start?

The important element for anyone thinking about targeting mobile customers is to start from the right point. Many make the error of starting by thinking that to target mobile customers, they simply need to build a mobile app. This is the wrong starting point.

You need to be clear why mobile is relevant and from the outset not get bogged down by tactical elements like what you are actually going to create.

Be clear about the basics: what are your goals, who are your customers, what will deliver value to your customers, what do you want your customers to do? In short: You need to know your audience.

For some brands, mobile apps are the right way to go, for many others mobile apps are a dead-end street.

It is important not to get drawn by the "hype" and "cool" of mobile, as this will almost certainly impact your ability to deliver value and reap a return from your effort to develop and distribute a mobile app in the first place. Not everyone uses apps; not everyone has an iPhone. It is important to create a service that will deliver value to all your customers and your business. You need to deliver a great experience to keep them all engaged.

Ultimately, you can develop great apps that augment your brand in new and innovative ways. From hairdressers, to garages, to independent fast food outlets, all could benefit in some shape or form from developing an app.

Restaurants can design an app to cash in on the features that resonate with customers most: information on menus and nutrition value. Hairdressers could integrate an app with their booking systems, enabling people to see the spaces free for their haircut. The salon could then incorporate a loyalty scheme and give discounts to people who book through the app. Linking in social feeds could also encourage the salon's customers to recommend to their friends, and get rewarded for it.

Garages could develop apps incorporating their skills and knowledge to provide more of a utility app for consumers, giving top hints and tips on how to identify an issue prior to taking their car into the garage. Again, this provides consumers with a reason to keep revisiting the app.

Building a Mobile Strategy

Mobile apps can do a lot. They can engage customers, drive footfall, encourage commerce, and deliver customer service. At the other end of the spectrum, mobile apps are also about entertainment and gaming.

How do you find the middle ground between what apps offer and your own business objective? Keep in mind that the role of mobile—as a medium—and apps as a channel to your customer can be divided into three groups:

- Save time (utility, such as apps to help customers book restaurants or tickets on the spot)
- Kill time (fun, including games, novelty, and music)
- Prime time ("wow" factor apps that also allow users to enjoy premium content or cool experiences)

Once you are clear on what you are setting out to achieve, you need to develop a strategy that will guide your mobile activities, steer and help prioritize your investment, and help you measure your results. After all, this is a business.

If your ambition is to engage customers, you need to create services your audience can use and appreciate. In other words, you have to be relevant and offer an experience that encourages frequent use. It's frequency that encourages engagement, boosts loyalty, and, ultimately, builds trust. It's a given that your mobile app should work on a wide range of mobile devices.

If you want your customers to buy products and services using your app, then it's crucial to optimize for mobile. There's no point trying to plug in elements designed for a PC experience—such as a check-out process—into your app. Your customers will simply not use them. And you will almost certainly lose the sale. And don't forget, reducing the number of keystrokes—or clicks/scrolls—is key when it comes to delivering a good customer experience and increases the likelihood the customer will actually complete a transaction. If a customer has to jump over hurdles to use your app, they won't return.

ALERT

People are relying on their mobile phones—and also mobile apps—at every stage of the consumer journey. From researching products, to conducting transactions, to sharing a product review with their social network, people rely on their apps to access advice and make the right decisions.

Make Them an Offer

With the avalanche of apps currently in the app market, it's no wonder that brands, businesses, and even bloggers are turning their attention to apps.

Clearly, apps, like any other channel to your customers, must match your audience demographics and demands. But delivering a great experience doesn't require you to make an all-singing, all-dancing app, chock-full of features and functionality. Think about your business and how people interact with you in real life first and then work back from there. That will help you identify, and integrate, the features that matter most.

Media

If you're a publisher, make sure your app offers the ability for users to cache and store content for reading offline. For users with downtime—on the way to work or during the day—this feature ensures that they download your content and read at a time that suits them best.

Shopping

Whether you make and sell fashion accessories, offer specialty sporting goods, or just have a secondhand shop, make sure your mobile app enables customers to complete their shopping activity over several visits and keep a list of items for later. Don't forget that shopping has become a social experience. Provide features that allow users to share images with their friends and family and connect to relevant reviews. A one-click to post to favorite social networks such as Facebook is a real plus.

Travel and Tourism

For brands where loyalty is important, such as a hotel chain, a mobile app should provide customers a simple way to manage membership accounts, store preferences, and sign up for loyalty benefits and offers they earn from their stay.

Finance

Whether its trading or financial services, your app must provide a way for customers to manage their accounts and activity and have a single view of all their services in one place. It is important to update, refresh, and add new elements to your app on a regular basis. Adding more value and interesting features will keep people coming back. Failing to do so will typically lose people's interest very quickly.

You should always aim to build a service based on a core foundation. Think about what your core product or service is and ensure that you get that right. Add features and value-added elements on top of your core service. You need to get your basic service right across all the mobile devices that your customers use. Failing to do so will limit your audience and will instantly lose a potential customer.

If your service does not work on their device, your ability to acquire customers will be instantly impacted. As well as thinking about acquiring customers, you should think about retention, what is it going to take to keep customers coming back? You need to keep all the customers you can.

There are certain features that will govern the approach and type of services that are required, and the likely investment. If there is a need for customers to be able to download content to read offline, use features like tilt and shake (accelerometer), use pinpoint location accuracy (GPS), or send content via text to people in their contacts book, you are going to be forced down the mobile app route. For most other features mobile web can and will deliver.

Once you have defined your strategy this should lead you to the right approach for you to take your service to your customers. For many, a mobile web strategy will be the right way to go; for others, a hybrid approach that combines mobile web and mobile apps or even pure mobile apps strategy may be right.

Remember, once you have lost a user it is extremely difficult to get that person back. For this reason, retention should be a key part of your thinking and strategy.

The App Environment

A good way to think about the mobile app environment is to break it down into two core areas; development and distribution.

Development

There typically is an in-house or external choice. Unless you have true expertise in-house, an external or third-party route is the best way and most efficient development method. Although the perceived costs may be higher, the expertise that a third party can bring should ensure that you don't run the risk of getting caught by the many pitfalls of mobile development.

ESSENTIAL

If you are using an external resource or third party to develop your mobile app, it is common that the process of managing the submission process of your mobile app to the various app stores is included in the cost.

If you are going down the in-house route, you will need to establish developer status with the various platforms you will be developing apps for. This can sometimes take time and cost money. Once established as a developer, you will need to have access to all the technical knowledge or Software Development Kits (SDKs) to be able to potentially develop apps for the individual platforms and devices.

If you are choosing a third-party supplier, always ensure that they have the skills and experience to deliver across the range of devices, platforms, and different versions of operating systems for which you want to offer services. Ask to see examples of their work that closely match what you are looking to do.

If you wish to develop for iPhone, make sure they fully understand the implications of all the different operating system versions, multiple devices, and platforms. Make sure they have had experience successfully porting services across multiple devices and platforms. If you wish to target BlackBerry business users, make sure they have the proven skills to deliver. Don't let them learn on your project. An experienced development partner will minimize the risk, especially when it comes to performance and submission to distribution partners.

Distribution

The next challenge is getting your mobile app in front of customers, downloaded, and used. Apps are typically distributed via a third party. These are either operated by device manufacturers such as Apple, Black-Berry, Samsung, and Nokia; operating system providers such as Apple (iOS), Google (Android), and Microsoft (Windows Phone); or third-party independents such as GetJar and Amazon.

ALERT

Make sure you get your app in front of people. If they can't find it, people will not download your mobile app. You will fall at the first hurdle.

Once your mobile app has been developed, it will need to be submitted to the distribution partner for approval. This process can take anywhere from a few days to several weeks. Make sure you factor into your schedule the potential time that this process can take. Also consider that there is a risk that an app is not accepted once it has been submitted. If this happens it will require the issues to be fixed and the app resubmitted for approval. This adds time to the approval process.

In the end, a mobile app needs to deliver value that can't be replicated through the mobile web environment. This could be through, for example, customization or personalization, use of the advanced device features, presentation of highly graphical content, the option to save content, or the ability to share content via text message and social media with contacts. This value should be able to justify the investment in developing a mobile app.

Sections of Chapter 1 contributed by Martin Wilson, Mobileweb Company, and Jennifer Hiley.

CHAPTER 2

App Market by the Numbers

Years ago, when the mobile market first started gaining momentum, no one could have guessed just how successful apps would become. Mobile has come a long way. Years ago, the only interactions consumers had with mobile were SMS messages and purchased ringtones, themes, and graphics for their phones that had limited functionality. Now it seems there is an app for everything, from games to books to navigation. With the market so overly saturated, it's important to know the numbers, because to make it in this app world, you need to understand the opportunities and know your competitors.

The Numbers Game

The mobile phone may have begun as a communications tool, but a raft of research shows that people are now using them to find, download, and enjoy apps. Indeed, a recent report by the consumer research and measurement firm ComScore (*www.comscore.com*), found that people are using their devices to do much more than talk, text, or browse the web. There is a new and intense focus on mobile app usage, where approximately 50 percent of consumers now use their phone to download apps.

With numbers like these, it is clear that the mobile ecosystem is growing rapidly, and it seems like there is a new rival on the scene every day. Whether you are planning on getting into the app market or you are already in it, it's important to watch high-level trends. Focus your time on tracking larger trends across the app stores, rather than looking for small patterns. Keep an eye on the changing tides of consumer behavior. A daily diet of important data will help you keep your thumb on the pulse of the market.

A good source that will help you track the state of the app market is Distimo (*www.distimo.com*). It offers a comprehensive overview and tallies how many apps are available across the major app stores. Here is some of the information you are likely to see:

- Apple: As of June 22, 2012, Apple offers 457, 987 apps to its iPhone customers and 155,708 iPad apps to its customers.
- BlackBerry: BlackBerry offers 90,000 apps to its customers.
- Google Play: As of May 2, 2012, Google Play offers 500,000 apps.
- Microsoft–Windows Marketplace: Microsoft offers 100,000 apps to its customers.
- Nokia: Nokia delivers 116,583 apps to its customers.

Number of Downloads

According to *Future Business Models and Ecosystem Analysis 2012–2016*, a recent report from research firm Juniper Research (*www.juniperresearch.com*), the mobile app market is in a "boom" phase. Specifically, the report found that consumers will download more than 66 billion mobile applications per year by 2016, more than double the 31 billion apps installed in 2011. That's a huge increase!

In addition, smartphones will continue to drive the majority of mobile downloads, although tablets, such as Apple's iPad, will yield one in every four app installs by 2016. Free apps are expected to make up 87 percent of all downloads. In this situation, postdownload activities like in-app virtual goods purchases and subscription sales will drive the revenue. Of course, games will remain the most popular app category throughout the forecast period, followed by multimedia.

App Genres

As in all businesses, you need to know what sells in the app market in order to make a successful app. In the case of apps, it's important to track the main genres of apps consumers are downloading and enjoying. There are a number of genres heating up in the mobile app marketplace, and where there is activity, there is opportunity.

It's no surprise that the apps that make customers' daily lives better and more efficient when they are out and about are found in the top grossing and top downloaded genres: Travel, Social Networking, and Games. Entertainment for customers who are on the go is also a top genre, including sports, music, photography, and video.

FACT

According to the analytics firm Flurry (*www.flurry.com*), mobile apps are gaining ground over web browsing among U.S. smartphone users, measured by the average number of minutes consumers spend with each per day. Specifically, consumers spend ninety-four minutes per day with apps, and seventy-two minutes per day using the browser. Flurry concludes that users appear to be switching to apps from browsers to access information, which may be the more convenient option through the day.

Apps over the Web

Because of the advancement in innovation across the mobile app landscape, there are a number of categories of apps that have evolved over the last twelve months, including entertainment, productivity, and education. Within these three categories, there are a significant number of subcategories for apps that consumers can adopt. In the category of entertainment, mobile apps now provide users with:

- Games
- Music
- Photography
- Social networking
- Sports
- Travel

In regard to productivity, mobile apps provide services related to:

- Business
- Finance
- Navigation
- Tools
- Utilities

Lastly, for mobile technology users looking to educate themselves through apps, mobile apps now show up in such categories as:

- Books
- Health care
- Fitness
- Lifestyle
- Medical
- News
- Reference
- Weather

Downloads Lose Luster

Are downloads a measure of success? Some research questions whether a singular focus on this performance indicator is a good idea. Indeed, *iPhone AppStore Secrets*, a study from Pinch Media (*www.crunchbase.com/company/pinch-media*) that analyzed over 30 million downloads from

Apple's App Store, reports that just 30 percent of people who buy an iPhone application actually use it the day after it was purchased.

And the numbers plunge from there. After twenty days, less than 5 percent of those who downloaded an application are actively using it. Of smartphone owners, 68 percent open only five or fewer apps at least once a week, finds a survey by the Pew Research Center's Internet and American Life Project (*www.pewInternet.org*). Seventeen percent don't use any apps, and about 42 percent of all U.S. adults have phones with apps, Pew estimates.

When you connect the dots, it's clear that it is not enough to simply get one app download or sale; you have to engage your customers and encourage interaction.

ALERT

Encourage your customers to be your app advocates. Provide them the tools to share, tweet, "like" and praise your app and you will be richly rewarded.

Mobilewalla (*www.mobilewalla.com*), an app analytic firm, argues outright that the number of downloads is a very poor measure of how popular an app is. They estimate that a whopping 80 to 90 percent of apps are eventually deleted.

App Interaction on the Rise

In addition to opening up new opportunities for businesses, this rapid growth of mobile apps and the mobile marketplace has shaped consumer behavior in terms of how they manage their daily lives. Not only do people want to use apps to connect with whomever they want, whenever they want, they also want to be able to manage their lives through apps on their mobile devices.

According to the Nielsen Company (*www.nielsen.com*), U.S. Android and iOS app users spent 101 billion minutes per month with their apps in March 2012, more than double the amount from a year earlier. This sort of finding prompts two key questions: Where are people getting the time to spend with their favorite apps? And what does this shift mean to your business?

Nielsen offers some answers. To start, apps are similar to other types of media that consumers engage with, and while there are more and more options, the amount of time available in each day remains the same. At the same time, consumers are devoting more time to certain apps.

FACT

In the first quarter of 2012, Nielsen reported the top fifty apps ate up 58 percent of U.S. users' app time. That figure was down from 74 percent in 2011, indicating there are new apps entering the market, as well as opportunities for new apps to capture consumers' attention.

With the amount of time spent engaging with mobile apps doubling each year, you can anticipate that this will open up significant opportunities for business owners, large and small.

World Becoming Public, Instant, and Global

The advance of social media has transformed how people communicate with one another—and with the brands, companies, and institutions they interact with on a regular basis. Marketing, advertising, and commerce have become conversations and it's little wonder that people increasingly rely on dialogues taking place on Facebook pages and via Twitter feeds for information and advice. Against this backdrop, the mobile app is not merely another way people can consume content: It's also another communications channel that encourages engagement—and can kick-start a conversation—between people and companies. The result is a new consumer trend everyone who owns a business needs to pay attention to. The world is becoming public, instant, and global, and smart app developers and brands will factor this shift into everything they do.

The advent of mobile devices and immediate access to data at any time has allowed everyone to share their daily activities with the public. Being able to tell your friends what you had for breakfast or how long you've been stuck in traffic with just a few taps on your smartphone is now the new norm. This is facilitated by access to a device that will share these communications, such as pictures, videos, and tweets, across a global network. This means that it is now possible for you to transmit the details of your personal daily life to the 6 billion people with mobile phone subscriptions in the world today. Someone halfway across the Earth can see your pictures, tweets, or videos if you want them to.

Even more, with the GSMA's (Global System for Mobile Communication) SS7 roaming network, which allows you to connect even on the go, and through the Internet, consumers now have access to all global subscribers in real time. This open framework has transformed consumers' daily lives and ability to communicate and share with people in and out of their network across the globe.

The key is immediacy, on both sides of the content. Mobile networks are now faster than ever and will continue to grow. These large global networks provide consumers with the ability to have instant access to the data and information that they seek. Not only can they pull information into their own private networks, they are able to immediately push the data and information that is most interesting to them back out into the network in a millisecond.

Hot Consumer Trends

Since these new consumer behaviors are evolving, the way businesses approach consumers must change as well. It is important to be cognizant of your audience and the trends they participate in. A few key trends are:

- **Continuous connectivity:** Because people have instant access to data and information, it is opening up a new breed of consumer trends. It is increasing productivity but also improving communication methods through social networking, tweets, instant messaging, advanced text messaging, and more. Therefore, think about including features in your app that allow your customers to connect and share.
- **Technology-based decision making:** Consumers are now able to leverage technology to make important decisions based on data available within their mobile apps. From travel to weather and news, consumers are able to find what they need in an instant. This ease of use has transformed their decision-making processes.

Now consumers merely turn on their phones and open appropriate apps to get the information they need. If this is your customer demographic, then architect your app to satisfy consumers' need for instant everything.

- **Time management shift:** This ability to find the information they need immediately has opened up a shift in the way people manage their time. Because people can find the information they are looking for more quickly and communicate more easily with people around the globe, they have more time to do other tasks. Think utility and make sure your customers can get things done on the move.
- **Location-based opportunities:** Because consumers have the information they need relative to where they are, this opens up new opportunities for businesses to target consumers in a relevant and secure way. The GPS unit within each device now enables businesses to identify and target consumers in a timely manner, so don't ignore this capability when you map out your app.

Staying On Top

The apps that work really do have staying power. For example, Android phone users spend about ninety minutes a day on their phone, about two-thirds of that on apps, according to media research firm Nielsen. This means that when you target your audience with an app they appreciate, you will be richly rewarded.

ESSENTIAL

With the advent of iPads and tablets, the mobile app game has completely changed. What we once called a smartphone has now evolved to take on a variety of forms. With this change in format, consumers and business users alike now have even easier access to mobile data, shopping, apps, and engagement. The mobile industry is seeing substantial adoption of these new devices, namely the iPad.

Nielsen estimates approximately 50 percent of consumers now use their phone to download apps. Around 33 percent used them for gaming,

27 percent for listening to music, and about 36 percent to access social networks.

So how do you get their attention? By making the perfect match between your audience and your app. Sure, some of it is the simple customer segmentation you know from basic business books. But you have to factor an extremely important variable into the equation: mobile.

Will Your App Stack Up?

You might be asking yourself, why do I need to know these numbers? How do they affect my app? It important to stay on top of market data and consumer research so you can keep up with your audience, but you also need to monitor what other companies and developers are doing so you can stay on top of your game.

Phase 1: Idea Exploration

So you've got a great idea for an app that's a match with your audience and what they require in an app. But before you ring up an app developer, or start coding it yourself, check out the market and see if your amazing new app idea holds any water.

Start by searching out competitors to see who the players are. Chances are if the developers of a similar app are also one of the industry giants, then you might just want to tuck that idea away for a while (or until you have a kajillion-dollar marketing and advertising budget to go head-to-head with those 800-pound gorillas). However, if your competitive landscape is populated by just a handful of independent developers, then you can feel free to proceed with your idea.

The next step is to download some of the apps that are offered by your nearest competitors and examine them with a fine-toothed comb. A road test will reveal what makes them tick and uncover shortcomings and areas where you can improve the experience and hence differentiate your app.

With your homework done, you'll need to drill down even further and check out the customer reviews. This means a glance over the five-star reviews, but make sure to pay even closer attention to the one- and two-star reviews. Ultimately, those people who are not so sold on your competitor's app are your best source of intelligence. Think of them as your own private (and unpaid) focus group. In their reviews, they are already telling you what works and what doesn't, and identifying the features they miss most.

This is great insight that money can't buy. What's more, this is the group of people you can target when your app hits the market. Since you built it with them in mind in the first place, you can bet they'll jump ship, hop on board with your better new and improved version of the app, and say good-bye to the competition.

ESSENTIAL

You don't have to download and road-test every app. You can study competitor apps you didn't download by checking out the screen-shots, graphics, and other details around the apps. Chances are the app developer has uploaded several images, which should give you a pretty good indication of the look and feel of the app. You'll then know where the bar has been set, and you'll know how to proceed when fleshing out your own app.

Now that you're finished exploring the competition, take out a notepad and jot down all of the things you think you can do better, record all of the gripes and complaints from dissatisfied users, and make notes on how you can set yourself apart graphically. Then, it's time to validate your idea.

Phase 2: Market Validation

The next market research tool to consider is the App Chart, which can be found in leading app stores and on resources like TopAppCharts.com (*www.TopAppCharts.com*) and AppFigures.com (*www.appfigures.com*).

What should you be looking for? Simple. Check and see where your competitors are on the charts. Are they top twenty-five? Fifty? Are they lost in the great app store abyss? These are all data points to consider when research-ing whether your app idea will be a success or not.

Imagine you found an app that's similar to your new app idea, and it's ranked in the top twenty-five in a particular category and isn't from one of the big studios. Congratulate yourself! You are obviously on to something, and you're one step closer to validating your idea.

Sadly, if the app that epitomizes your idea is at the bottom of the list, then that's a sure sign that your app is not a crowd pleaser. At the end of the day, creating apps is a business. And, as with any successful business, economics plays a key role. You need to ask yourself, "Will my app idea make money? And if so, how much?"

There is only one way to find out: Conduct more market research!

Phase 3: Revenue Projection

The good news is this part of your app journey isn't as difficult as you might think. Sure, you can dig deep in the numbers and the forecast and run the software to build a business plan, but you can also get the answers you need by learning to listen to the marketplace.

During a recent podcast interview between AppClover.com (*www.App Clover.com*) and Taylor Pierce, author of *Appreneur*, Taylor was asked if there is any way to predict potential revenue of an app idea in order to determine whether it's worth moving forward on. Taylor answered, "Yeah, just go into forums and ask."

Wait a minute! That's a revenue projection tactic? Yes, it works. No reason to make it more complicated. No software to learn, no scripts to run, no analytics to study. Just engage with the community. Here's how Taylor suggests you do it:

Say for instance, the competitor's app that you're thinking about creating is hanging around the top twenty-five spot pretty consistently for a certain category. How much is the developer making on that particular app? You can find out by participating in a development forum (like iPhoneDevSDK .com, for example, if you're an Apple iOS developer) and asking other members—developers active in the same category with apps that are likewise ranked in the top twenty-five—what they are seeing in downloads and sales.

If they tell you they generate around 500 sales per day, then just do the math; the number you have allows you to forecast with a level of confidence how your similar, but new and improved app, will do when you launch. You can count on doing as well (if not better) than the other developers you asked.

In addition, while you're in the forums, you might as well ask app developers across other categories what they are also doing in terms of sales.

That way, when you come up with a great app idea in a different category, you've already got some of the revenue projection research finished.

Sections of this chapter were contributed by Matt Lutz, AppClover.

CHAPTER 3

Know Your Audience

The old rules of marketing, affectionately called the Four P's, still apply today in the new world of mobile. In order to stay relevant, you need to expand on the old ways of thinking. The new rules today also call for precise segmenting and targeting, so you can add one more P to the mix—People—and make it the Five P's: Product, Price, Place, Promotion, and People.

Knowing your audience is absolutely paramount in the mobile app world. Gone are the days of blindly marketing your product without specific targeted goals and metrics. If you don't know your audience, you are dead in the water in the mobile app landscape. How do you get to know your audience? By segmenting the mobile market across the top genres and pinpointing your particular app's focus to those genres.

Segmenting the Market

As you look at the mobile app landscape, it is critical to segment the global audience across the following age groups:

- 8–12 Constantly Connected
- 13–17 Digital Natives
- 18–24 Generation M (for Mobile)
- 25–34 Millennials or Gen-Y
- 35–54 Gen-X
- 55+ Baby Boomers

To understand your audience even deeper, you can match the demographics with smartphone use. According to eMarketer (*www.emarketer .com*), smartphone ownership by age in January 2012 was as follows:

- 18- to 24-year-olds: 62 percent of all subscribers
- 25- to 34-year-olds: 66 percent of all subscribers
- 35- to 44-year-olds: 58 percent of all subscribers
- 45- to 54-year-olds: 45 percent of all subscribers
- 55- to 64-year-olds: 33 percent of all subscribers
- 65-year-olds and over: 22 percent of all subscribers

Mobile Demographics (Ages 8–34)

Every age group interacts with technology in a different way. Members of one group may have grown up in a time when mobile devices were readily available, while members of another group remember a time when using a landline was the norm. When you have to decide which age group to focus your marketing efforts on, it is important to first understand what these groups expect from their mobile technologies.

Constantly Connected

We've all seen it: the toddler that knows how to unlock her parents' iPad, open up her favorite game, and start playing. The days of plastic squeaky chew toys as a method of entertaining children is gone. They want digital!

Eventually, the youth of this country won't even need a computer or laptop; everything will be functional from one single handheld device. It is important to watch how the children of today use mobile in their everyday lives in order to better build apps specifically targeted to them as they grow older.

Interestingly, this also appears to be a demographic weaned on Apple devices and apps. A study by the NPD Group entitled *Kids and Apps: A New Era of Play*, reports that the Apple App Store has become the go-to destination for parents to find apps for their children. A full 62 percent of parents conduct searches on the App Store and more than two-thirds of parents surveyed look for age-appropriate content. For parents with kids ages 2–5, 75 percent search for age-appropriate apps.

Nielsen numbers show that tots are truly tuned into mobile devices, using them for both education and entertainment. Read between the lines, and there is a huge opportunity for app makers to develop apps that specifically target these cyber children. But the research also cautions developers not to ignore the tablet. The bigger screen size of the tablet is better suited for a child than a mobile phone and is portable enough for parents to use as a quick fix to keep kids busy on the go.

The NPD Group study also revealed that gaming apps are the most popular for boys, who play an average of 10.8 hours per week. Girls, on the other hand, are more interested in creative apps (music, art, and photo) and spend 9.5 hours a week on social networking sites. Kids ages 2–8 use educational game apps, while the tween/teen set (12–14) lean on social networking and music apps.

Digital Natives: Ages 13–17

This generation was born into and raised in the digital world. The economics, politics, culture, and even the definition of family will transform and evolve for this generation. This demographic, made up of people who were weaned on the Internet, cannot imagine a moment without their mobile phones. It's a dependence documented by reams of recent consumer surveys. For example, one Digital Native famously said: "I'd rather give up my kidney than my phone."

Of course, it can't be verified that comments such as this are representative of the entire population of Digital Natives. However, with mobile phones becoming ubiquitous and smartphone penetration on the rise, it's obvious that mobile devices—and increasingly apps—have become a constant companion for Digital Natives in all scenarios. This speaks volumes about the mindset of this demographic and what you need to factor into your app to be a crowd pleaser.

With this group, sharing is essential. Make sure your app meets the basic requirement of Digital Natives: to connect. Mobile social media communication is more than a pastime; it has become an obsession. Research shows Digital Natives connect with their communities at all times and in all places: when they get up in the morning, before they go to bed at night, while they are in the bathroom.

ESSENTIAL

The real revolution here is not in the breadth and types of mobile apps Digital Natives can access using their smartphones; it's in the new channel to the customer it gives developers and companies that created the apps in the first place. Clever app developers and companies will understand the importance of user input and encourage it. And truly cutting-edge companies will build user participation into all aspects of what they do.

Marketing isn't selling. With Digital Natives, marketing is all about engagement. Whether it's offering new ways for them to communicate with their peers and social groups such as Instagram, Path, Highlight, or Pinterest; making their daily lives easier with Astrid, Glympse, or Stumbleupon; allowing them to purchase items more easily through Square, Google Wallet, or LevelUp; or providing them the opportunity to co-create the brand message they want, really listening to what Digital Natives have to say can clinch the deal.

Generation M (for Mobile): Ages 18–24

Generation M grew up posting every possible picture throughout their lives to Facebook, tweeting every movement they make, sending thousands of SMS messages a day, sharing videos of their daily events, never afraid to publicize every aspect of their lives. This generation is paving the way for the Digital Natives and Constantly Connected Generations. Because of Generation M, the age groups that follow in their footsteps will be faster, quicker, more agile, and, some could argue, more innovative.

Millennials, Gen-Y, Start-Up Generation: Ages 25–34

The 77 million people in Gen-Y, or the Millennial Generation, have embraced and advanced innovation across the mobile landscape. This generation is addicted to creating new things and is now leading the start-up revolution found in the United States. This is a generation of entrepreneurs who understand what it takes to successfully start their own businesses. Forty percent of this generation envision starting their own business, and about 20 percent already have, according to a recent report published by the Affluence Collaborative, a research partnership.

This generation lives and breathes mobile. They have watched the advancement from feature phones to smartphones, the movement from texting to the evolution of mobile apps. These young adults are now leading the innovation across mobile apps. Everything about this generation is mobile, including gaming, social media, interacting with groups and crowds, crowdsourcing, commerce, information gathering, and family planning.

Gen-X: Ages 35–54

Generation X is an estimated 50 million strong, and is a highly educated and sophisticated group, with more than 60 percent of the population having attended higher education institutions. Generation Y consumers have grown up with technology and smartphones, but Gen X represents those consumers who remember a time without digital advertising. Therefore, they respond equally well to modern and traditional marketing.

While Boomers (think Steve Jobs and Bill Gates) helped push the first digital revolution, Gen X (think Google's Sergey Brin) is picking up where they left off and advancing the web into new areas never before imagined possible. Gen Xers are more educated than their Boomer elders and are putting that knowledge to good use.

This generation helped start the evolution of mobile. This generation includes a segment of highly affluent consumers who use mobile for activities such as mobile banking, mobile commerce, travel activities, checking news, weather and sports scores, and living a daily mobile life.

Digital Moms

Digital Moms reach out to mobile during every stage of the purchase funnel. They want to research purchases and make smart and informed shopping choices. As a whole, this mobile-savvy demographic is earning, spending, and influencing spending at a greater rate than ever before. Developers and brands would do well to focus their efforts and market more effectively to this growing and connected customer segment.

Digital Moms in the United States are accumulating more wealth, making more shopping choices, and using more technology than the average person. Digital Moms are more likely to be owners of smartphones and more likely to consult others when making purchase decisions. If you are creating an app for your business, this demographic is important to focus on. How should you tap into this important segment? Here are some statistics to help you on your way.

The Stats

Moms have money. An article from Asking Smarter Questions (*www .askingsmarterquestions.com*) provides some key statistics about both women and mothers. For example, moms represent a $2.4 trillion market in the United States. By 2028, the average American woman is expected to earn more than the average American male. Women account for $7 trillion in consumer and business spending in the United States. Women are making more than 85 percent of all consumer purchases, including more than 50 percent of traditional male products, such as cars, power tools, and consumer electronics.

In addition, moms are socially connected via mobile. In a study by NM Incite (*www.nmincite.com*), at least half of the moms surveyed access social media from their mobile device. In an article from Asking Smarter Questions, 64 percent of moms ask other mothers for advice before they purchase a new product, and 63 percent of all mothers surveyed consider other moms the most credible experts when they have questions. Additionally, 92 percent pass along information about deals or finds to others. Moms are using Facebook and Pinterest to share tips, product recommendations, and shopping deals with their friends.

Moms are savvy and connected shoppers. In fact, moms are 70 percent more likely to download online coupons from retail marketing websites and 65 percent more likely to download coupons from a manufacturer's website than the average person. Additionally, 92 percent of moms pass along information about shopping deals or finds to others. Fifty-five percent of moms who use social media daily said they made their purchase because of a recommendation from a personal review blog. Almost 20 million online moms will read blogs and one in three bloggers *are* moms. Of those mommy bloggers, 77 percent will only write about products or brands whose reputations they approve of, and another 14 percent will write about brands or products they boycott.

Moms are demanding mobile users, to say the least! Nielsen notes that some 54 percent of American moms are using smartphones. Considering 50 percent of the United States population owns a smartphone, mothers are more likely to be smartphone owners than any other demographic. Additionally, moms are about twice as likely to own either a smartphone or a tablet compared to other women.

Not all digital moms are the same. Advertising Agency MWW surveyed 1,000 moms to better understand their use of digital and social media. They found five digital mom archetypes: Mobilizers (younger, hyper-connected moms using mobile), Urban Originals (influential tastemakers), Practical Adopters (working moms who use technology to stay ahead), Casual Connectors (older moms who use mobile technology to connect with their kids), and Wallflowers (younger moms who are consumers of media). Providing multiple means to connect with these subgroups is important.

Moms are also more likely to shop online. Women account for 58 percent of all total online spending and 22 percent shop online at least once a day according to Asking Smarter Questions. Additionally, when using social media, moms are 38 percent more likely to become a fan of or follow a brand online, and moms who blog are more than twice as likely to follow brands and celebrities compared to the online average. Clearly, Digital Moms are mobile-equipped, multitasking, socially connected, family-protecting, brand-loyal, savvy shoppers. They are well-informed consumers who share what they know with other mothers.

What are the keys to tapping into the Digital Moms customer segment with your app? Be relevant and authentic. Meet them online where they are. Provide moms with tools, lists, and apps to help them manage their lives. Take the time to understand their needs, build their trust, and deliver on your promises and you just might be lucky enough to have a new army of moms who come back for your app.

Boomers Are the Boom Market

Mobile app developers have a huge opportunity to tap into the giant wallets of the 78 million Americans who call themselves Baby Boomers. Boomers are Americans born between 1948 and 1964, and they make up the largest customer segment in the United States.

Indeed, boomers are a *big* group with a lot of buying power to match. They control 77 percent of the nation's wealth, and they buy 45 percent of all consumer goods. In fact, Boomers spent $2.5 trillion in 2010 alone. And, since they are all getting older, this is a market that shows no signs of slowing. There will be an additional 53+ million people over fifty in the next twenty years.

And forget the stereotype that Boomers are Luddites when it comes to mobile technology. The Pew Internet Project revealed that 80 percent of Boomers are online and 46 percent of them own smartphones. They spend fifteen hours per week online doing research, e-mailing, shopping, reading, and socializing about their hobbies like gardening and travel. Mobile phones are quickly becoming the connection of choice for Boomers.

Boomer Survey Surprises

According to a national online Mitchell Poll of 600 Boomer smartphone users in June 2012, 76 percent have downloaded at least one app in addition to standard apps. Nearly half of all the Boomer respondents (49 percent) have added six or more apps to their smartphones.

The survey also found that Boomers love text messaging. The following are some surprising statistics:

- 88 percent use text messaging
- 52 percent text one to five times a day
- 20 percent text six to ten times a day
- 16 percent text more than ten times a day

It's a known fact that Facebook is the social medium of choice for Boomers, who love to connect with old friends and relatives throughout the country from their home computers, but they are also enjoying the social activity from their smartphones as well. The Mitchell Poll found that 51 percent use Facebook on their smartphones at least once a day and 41 percent log in one to five times a day.

The Mitchell Poll also found that more is better when it comes to the number of apps on a Boomer's smartphone. In fact, 22 percent have added six to ten apps; 10 percent have added eleven to fifteen apps; and 17 percent have added more than fifteen apps.

Apps That Appeal

According to AARP, nearly 70 percent of Boomers use the Internet to shop and peruse travel deals. According to Drug Store News, 52 percent of Boomers 50–54 years old purchase health and beauty aids online, which is more than any other demographic.

In short, this group is primed for easy-to-use shopping and travel apps.

When it comes to health care, Boomers are especially interested in apps that help manage or ward off chronic disease, the Mitchell Poll discovered. This is the segment to target if you want to offer health care or wellness apps. In fact, 57 percent surveyed by the Mitchell Poll said they would likely download a general information app to look up medical symptoms or diseases. Additionally, 48 percent would likely download a health monitoring app for a specific chronic medical condition, and 48 percent would likely download a weight and exercise app.

Boomers are willing to purchase health and wellness apps as well. The Mitchell Poll indicates that 71 percent would spend at least a $1 on a health and wellness app and 50 percent would spend at least $2 on a health-related app.

FACT

There is also a future in games for app developers who want to target Boomer women, particularly in casino and health care games. About 40 percent of gamers are women and nearly one-quarter are over fifty.

Understanding the Boomer Psyche

Developers and marketers need to better understand the mentality of Boomers to capture them as customers.

Boomers have a unique historical perspective about technology. They were in their twenties and thirties when the first IBM PCs and Apple computers appeared. They became empowered by this technology, an experience that made them the first real early adopters of that era.

But Boomers also lived much of their lives before the Internet, social media, online content, advertising, and all things digital.

Boomers recall a time before mobile when telephones had wires. They watched TV delivered by a handful of network broadcasting stations, not Hulu. And when their friends moved to another state they lost touch; there was no way to connect via Skype or Facebook.

You should also remember that Boomers' ideals were developed during the Vietnam War years, an era when human rights and individual freedoms were primary concerns. Their concerns about privacy and security continue to impact their attitude about technology today. According to a 2009 study conducted by AARP and Microsoft, 67 percent of people ages 18–24 said they were concerned about online privacy. The number increased to 86 percent for people 55 and older.

FACT

According to the Centers for Disease Control, 80 percent of adults over 65 in the United States have one chronic disease, such as diabetes or heart disease, and 50 percent have at least two. Boomers who want to help manage their chronic disease care are willing to pay for the mobile apps to help them, but they also need assistance from app makers in learning how to use the apps. Develop a program to help your app help Boomers and make them comfortable about using your app and you will increase app engagement and your customer's quality of life.

It is clear that Boomers want technology to fit into their lives and meet their values, but they also expect technology to adapt to them, not the other way around. *Age Wave* author Dr. Ken Dychtwald identifies four areas where Boomers will focus their quest for vitality:

- **Financial vitality:** Boomers will work to manage their money and maintain their assets, along with "helping" their children and grandchildren with college loans and homes.
- **Physical vitality:** This relates to both appearance and health and fitness. Boomers already spend countless billions trying to maintain vitality in these areas, and it will only increase in the coming years.
- **Mental vitality:** The impact of mental diseases like Alzheimer's have made a tremendous impact on today's Boomers, who have seen it in parents, grandparents, and loved ones. They will spend money, time, and energy to avoid a similar fate. Companies are capitalizing on this quest for mental strengthening by developing apps with series of games and puzzles.
- **Social vitality:** As a generation that created and managed social networks in the real world for the last forty to fifty years, Boomers will

want to maintain their vast social networks. They've invested too much of their lives in it, and they are not about to uproot themselves and relocate to some distant retirement community. That's the biggest reason Boomers will likely stay in their communities with their families.

How to Reach Boomers

If you are serious about getting something from the Boomers' $2.5 trillion wallets, here is some helpful advice.

- **Size matters:** Make your app clear and easy to read. You may want to use a thin font in a small point size that looks cool; however, it will be too hard for Boomer eyes to read. Use a sans serif font like Tahoma, Calibri, or Arial. And never use a font size less than 12 point.
- **Easy does it:** Light or bright text against a dark background is hard to read and even hurts the eyes. Try to limit reverse type to headlines only. Use a lot of white space. Primary colors are a plus as well.
- **Keep it simple:** Make navigation clear and simple. Use large buttons placed in areas that are easily recognizable. And don't make users search for the save, share, or help button.
- **Explain clearly:** Make the "help" section and home page accessible from every screen if possible. Boomers absolutely hate being lost in a maze with no way to access help or get back to the home page. Also, make sure your help section uses easy-to-understand terms in its FAQ, not technical jargon.
- **Healthy is better:** When it comes to health and wellness products, Boomers are *the* early adopters. Health and fitness mobile apps are a natural next step, provided developers deliver value and ease of use.

If you follow these simple steps, you are destined to delight Boomer users. Boomers naturally think of themselves as young and cool—after all, their ranks include dozens of pop icons ranging from Mick Jagger to Madonna. So stop thinking in stereotypes and make an effort to address this demographic. Remember Boomers want to stay relevant, and this also means having the latest tech toys and equipment. Developers who make the effort to understand Boomers will open their hearts, and their wallets.

You Are Mobile

People are mobile, and mobile is part of who they are. People's growing identification with and dependence on mobile devices was first captured in *Personal, Portable, Pedestrian* (2005, MIT Press), a milestone book written by Mizuko Ito, a renowned cultural anthropologist at Keio University in Japan, together with Daisuke Okabe and Misa Matsuda.

The book, released long before mobile apps became the norm, outlines the pivotal importance of the mobile phone, based on the fact that it is *personal* (you can customize and personalize your mobile devices and regard them as an extension of your personal identity), *portable* (you can have them on your person the majority of the time), and *pedestrian* (you can make them a part of your real-time life as it happens).

Fast forward, and the advance of apps has transformed mobile completely. Mobile was once a tool that allowed people to construct an intimate environment in which they could interact with friends and family members. Today, in a networked, wired-up world, mobile apps allow people to manage their daily lives, enjoy content, and engage in much larger conversations with an *extended family* of communities, organizations, companies, even governments.

The Tie That Binds

Clearly, people's lives and their devices have become inextricably intertwined. Mobile—and now increasingly mobile apps—empower people to capture and consume content. Apps impact how, when, and where people connect with friends and family; it assists in daily decision making; and it links physical and digital worlds.

Sections of this chapter were contributed by Suzie Mitchell, Mitchell PR; and Lisa Ciangiulli, Optism.

CHAPTER 4

You Are Not Alone

Building a mobile app can be somewhat of a daunting experience, but rest assured, you are not alone! There are endless resources available to guide you on your path. From helpful websites and global developer communities to local organizations assisting developers and brands as they navigate new terrain, you can reach out for advice and answers during every step of the journey. Don't be intimidated by the technology, the acronyms, the endless opportunities, or the expansive amount of resources available; the following sections cover everything you need to know to move your app strategy an important step forward.

Resources for Building

Once you've decided you want to build an app, you may feel immediately overwhelmed. There are so many different parts of the app process that you need to consider, how do you know where to begin? Figuring out the best plan of action in order to reach your app goal can feel impossible at times, but luckily, there are resources that can help get you started.

Websites for Reference

There are a number of excellent mobile industry resources, from small independent developers, to large organizations dedicated solely to mobile. The following list of websites will give you a starting point for analyzing the mobile market today:

- Canadian Wireless Telecommunications Association (*www.cwta.ca*): The Canadian Wireless Telecommunications Association (CWTA) is the authority on wireless issues, developments, and trends in Canada. It represents cellular, PCS, messaging, mobile radio, fixed wireless, and mobile satellite carriers as well as companies that develop and produce products and services for the industry.
- Consumer Electronics Association (*www.ce.org*): The Consumer Electronics Association (CEA) unites 2,000 companies within the consumer technology industry. Members tap into valuable members-only resources: market research, networking opportunities with business advocates and leaders, up-to-date educational programs and technical training, and exposure in promotional programs.
- CTIA—The Wireless Association (*www.ctia.org*): CTIA—The Wireless Association is an international nonprofit membership organization that has represented the wireless communications industry since 1984. Membership in the association includes wireless carriers and their suppliers, as well as providers and manufacturers of wireless data services and products.
- GSMA (*www.gsma.com*): The GSMA represents the interests of mobile operators worldwide. The GSMA unites nearly 800 mobile operators with more than 230 companies in the broader mobile ecosystem, including handset makers, software companies, equipment providers,

and Internet companies, as well as organizations in financial services, health care, media, transport, and utilities. The GSMA also produces events such as the Mobile World Congress and Mobile Asia Expo.

- Mobile Entertainment Forum (*www.mefmobile.org*): MEF is a global trade association for companies wishing to monetize their products and services via mobile. MEF is a member network with international reach and strong local representation.
- Mobile Marketing Association (*www.mmaglobal.com*): The Mobile Marketing Association (MMA) is a global nonprofit trade association representing all players in the mobile marketing value chain. With more than 700 member companies, the MMA's primary focus is to establish mobile as an indispensable part of marketing. The MMA works to promote, educate, measure, guide, and protect the mobile marketing industry worldwide.
- Mobile Monday (*www.mobilemonday.net*): Mobile Monday (MoMo) is an open community platform of mobile industry visionaries, developers, and influential individuals fostering brand-neutral cooperation and cross-border peer-to-peer business opportunities through live networking events to demo products, share ideas, and discuss trends from both local and global markets.
- Mobile Technology Association of Michigan (*http://mobiletechassnmi .wordpress.com*): The Mobile Technology Association of Michigan (MTAM) is a Michigan-based trade association for the mobile/wireless industry.
- My Wireless.org; America's Wireless Voice (*www.mywireless.org*): MyWireless.org is a national, nonpartisan, nonprofit consumer advocacy organization that gives wireless consumers the ability to protect the value and security they enjoy with wireless service.
- Open Mobile Alliance (*www.openmobilealliance.org*): OMA is an industry forum for developing market-driven, interoperable mobile service enablers. OMA was formed in June 2002 by approximately 200 companies, including mobile operators, device and network suppliers, information technology companies, and content and service providers.
- Wireless Industry Partnership (WIP) (*www.wipconnector.com*): Connecting mobile developers to information and offering partners from the wireless sector to build and showcase their products and services.

Strength in Numbers

There are a lot of tools and resources to help you build your app, but it is important to keep in mind that there are a lot of potential partners you can work with as well. By creating partnerships and alliances, and getting out into the community (and building up your own), you'll create a network of support and an army of supporters to help market your app.

FACT

Is Detroit possibly the next Silicon Valley? You bet! Thanks to its long history of automotive manufacturing and innovation Detroit—and Michigan overall—is poised to become the national hub for mobile development in several areas including telematics, a perfect fit with efforts by carmakers to introduce apps into the car. What's more, Detroit actually has the largest concentration of engineers anywhere in the United States. To match supply and demand, *Mobile Technology Association of Michigan* (MTAM) hosts "speed dating for businesses," bringing app developers together with companies that can utilize their services on a contract or employee basis.

One place to start is by connecting with your fellow developers. One of the best ways to drive discovery and downloads of your app is through cross-promotion within other apps. There are companies that will help you with this (for a fee), but by connecting with other developers, you can do no-cost cross-promotion trades, where you advertise the other's apps to your users within your app. This has two big advantages: By working with developers whose apps appeal to a similar target market as yours, you know you're getting the right eyeballs. Second, it's free traffic to your app, which you can then monetize through advertising, in-app purchases, or another freemium model.

Meeting Developers

Events are a great place to meet developers in person. Search Meetup .com (*www.meetup.com*) for developer groups in your area. You can also check out WIP's Mobile Community Calendar (*www.wipconnector.com/ events*) to find conferences, hackathons, and more. You can also look at sites like StackOverflow (*www.stackoverflow.com*), Quora (*www.quora.com*), LinkedIn (*www.linkedin.com*), Facebook (*www.facbook.com*), and Twitter (*www.twitter.com*) to meet developers online.

Events are also a great way to make connections with developer programs. Many companies such as operators, handset vendors, and tool providers have these programs in place. These companies often have promotional outlets and opportunities that can help get your app in front of potential users, whether through an app store of their own, advertising to their customers, or other promo slots.

For instance, companies that provide app developer's platforms, cross-platform development tools, or just community support are naturally eager to showcase just how developers are using their resources to achieve amazing results. To help developers spread the word these companies will also offer some promotional opportunities. OS and handset providers also have an interest in promoting apps on their platforms and devices. Understand that these won't happen for your app out of the blue; you'll need to build a relationship with these programs, and meeting their representatives at events is a way to start.

Hackathons

Hackathons are a great environment, too, because they not only allow you to get direct support from a resource provider, but also to show them what you're working on and catch their attention. Once you are there, don't worry so much about the prizes!

ESSENTIAL

Hackathons are evolving from an event that attracted app developers, to a happening that hosts a room full of social innovators—including designers, anthropologists, usability experts, business and civic leaders, and more. These stakeholders share one goal: to produce potentially groundbreaking ideas that power great apps.

By coming to hackathons with some ideas for apps and building a team to work on them, you can meet like-minded developers, or developers who have some experience in the same space that you're working in.

Bloggers and Press

Bloggers and the press are another great resource to use. You need to think outside the tech blogs, for the most part, and focus on the blogs your target users spend their time on. Are there app review sites they use? Or are there other niches you can explore? For instance, if you're building an educational app or app for kids, try to make connections with "mommy bloggers;" if your app is about soccer, find the forums and blogs where your users hang out and connect with them. Don't ignore local press, either: Many hometown newspapers and TV stations love to feature the work of local app developers (this is likely to be easier in smaller towns rather than Silicon Valley!).

You should also try to make other inroads in your local community through regional mobile/tech groups and associations. These groups often provide promo events such as demo days and start-up competitions, but also work with a lot of larger groups and events like CTIA and MWC (see this chapter), offering free passes and other exposure that you can take advantage of.

For example, the Mobile Technology Association of Michigan (MTAM), the only state-based nonprofit trade association for the mobile industry in the United States, hosts Mobile Matchup events for app developers and businesses and mobile developer training programs.

ALERT

What you are really trying to build for your app is momentum. Think of it as a snowball rolling down a hill—as it goes, it grabs more and more bits of snow, and picks up speed, getting bigger and faster. Each little bit of support you get adds to your snowball!

The App Economy Is Here!

The proliferation of connected devices and platforms has encouraged companies and developers everywhere to get in on the action. More apps may mean more choice for consumers, but it also turns up the pressure on everyone else—publishers, marketers, brands, businesses, and developers—to learn how to do business in a new and booming economy, widely referred to as the "App Economy."

Listen to Your Peers

Any way you look at it, the App Economy is going to be huge. Its growth is intertwined with the phenomenal increase in mobile phone penetration. To date, four out of five people on the planet now own a mobile phone. Moreover, smartphone ownership is accelerating even faster, accounting for 40 percent of devices shipped. In other words, sales of smartphones are outpacing sales of simple-feature phones. This is a significant shift because the rise of smartphones is also driving the growth of the App Economy. According to some estimates, there will be 36 billion apps downloaded in 2012!

This growth presents a huge opportunity for mobile developers to capitalize on their apps, but it also underlines the importance of understanding more about what the driving force behind the App Economy—developers—think about key issues such as platforms and payments. Knowing their choices will help you make yours.

ESSENTIAL

When you are picking partners and partnerships to power your app, don't forget the mobile operator. Once considered a "dumb pipe," many operators are getting smart and offering APIs that allow developers to quickly bake functionality such as messaging, billing, and location data into their app. Operator networks are good at sending notifications to people via text messaging, so it follows that operator APIs can help you incorporate SMS (or MMS—picture messaging) into your app in innovative ways.

Developer Economics 2012

Fortunately, *Developer Economics 2012*, now the third report in a milestone series, was created to map and define the new App Economy. This report is freely available for download (*www.DeveloperEconomics.com*), thanks to the sponsorship by BlueVia (*www.bluevia.com*)—the new global developer platform from Telefónica.

Drawing from an online survey of more than 1,500 developers from across the globe, as well as twenty qualitative interviews, the report sheds light on how developers in North America, Europe, Asia, Africa, Oceania, and Latin America view business opportunities and obstacles.

Among the insights:

- The average per-app revenue for a developer is in the range of $1,200–$3,900 per month, depending on platform.
- Irrespective of the platform they use now, the majority of developers (57 percent) plan to adopt Windows Phone.
- Apps for phones aren't enough now that tablets are becoming mainstream; more than half of developers are now targeting tablets.

What's the determining factor for developers when it comes to choosing a platform and building their business? *Developer Economics 2012* reports developers are motivated by a platform's reach across the number of devices. Significantly, reach is a far more important parameter for developers than either development costs or revenue potential.

Of course, these choices vary depending on the platform. For instance, Windows Phone developers are hardly motivated to adopt the platform because of its reach, since the reach of Windows Phone is still quite limited. They choose Windows Phone because development costs are low, and because it is a familiar development environment. On the other end of the spectrum, BlackBerry and iOS developers chose these platforms because of their large installed base and their revenue potential.

The Platform Race

Which are the hottest platforms right now? Android and iOS are running the show. In fact, Android and iOS have gone from strength to strength, being used by 76 percent and 66 percent of developers surveyed in the report by Vision Mobile respectively, in 2012.

It's interesting to note that mobile web is the third most commonly used platform, despite losing some ground year-over-year. The massive influx of software developers to the mobile industry has greatly contributed to this trend, as mobile web apps are inherently platform agnostic. The decline of BlackBerry and Symbian comes as little surprise, as Nokia and RIM have been floundering in the past couple of years, despite their past dominance in the global smartphone market.

What's the next platform that developers will take up? Research suggests that Windows Phone is "the new cool." Over 55 percent of developers are thinking of adopting Microsoft's renovated platform, especially as anticipation over Windows Phone 8 approaches its peak.

Developer Intent

Developer Economics 2012, which also measures developer intent share, the percentage of developers planning to adopt each platform, shows that Windows Phone is a clear winner. Its intent share has grown from 32 percent in 2011 to 57 percent in 2012. Android seems to be in decline, but that's not really the case. Most developers are already using Android, so there's not much room for growth for the platform.

Windows Phone seems the obvious choice for developers. Most developers have already tried their hand at Android, and iOS developers are now looking for a complementary platform to extend their reach.

ESSENTIAL

Choice of platform is not the only issue for today's developers. There are a plethora of choices to be made! Should they focus on native or cross-platform development? Which is the best revenue model among the several available? In the end, there's no single answer. It's all about choosing the right strategy that fits your company, your strategy, your app, and your budget. In the app economy you have to spend money in order to make money!

The App Poverty Line

What can you expect to earn making apps? That's a tough question. The plain truth: Despite the many opportunities around apps that flood the mobile industry, most developers are hardly breaking even.

Monetization has always been a thorny issue for developers, although it has improved as app stores have become a primary route to the market, providing developers with immediate access to a wide audience of customers. Overall, what you earn depends a lot on the platform you choose. Interestingly, the report shows that OS and, surprisingly, BlackBerry are in the lead. Android and Windows Phone developers are making considerably less money.

How hard is it for developers to make real money with their apps? A clear indication comes from the *Developer Economics 2012* report. It reveals that an app has a 35 percent chance of generating $1–$500. Moreover, one in three developers lives below the "app poverty line," which means they cannot rely on apps as a sole source of income, even if developing multiple apps.

BlackBerry developers make more money on their apps because the app store, BlackBerry App World, is less crowded than the app emporiums run by the likes of Apple and Google. What's more, the apps in the BlackBerry App World largely appeal to an audience of business professionals. This customer segment—the one that made BlackBerry the device of choice for the enterprise—is mobile savvy and needs no convincing when it comes to buying apps.

Also a lucrative business for developers is iOS development (what Apple uses), especially when compared to Android development. In fact, iOS developers generate, on average, 37 percent more revenue per app-month than Android developers.

So, why does iOS beat Android in the monetization game hands down? This is linked to several factors, including its strong appeal to an affluent audience; compelling content; and a stockpile of blockbuster apps, dominance in the tablet market, and frictionless payment. Users of the iPhone have shown themselves willing to pay for apps in a way that Android users have not. In January 2012, Apple said that since 2008, when its App Store opened, developers had been paid a total of $4 billion, of which more than $700 million was paid in the last quarter of 2011 alone. Google hasn't given a comparable figure, though Horace Dediu, who runs the Asymco consultancy, puts the figure for Google's total app sales in 2011 at $300 million, meaning developers would get $210 million in total.

Reality Bytes

The Mobile App Economy is in full swing, but does it mark a new chapter in history? Can the mobile industry learn from history to make sure everyone in this new ecosystem—particularly independent app developers—benefits and prospers? Drawing comparisons between the current app frenzy and the California Gold Rush of 1848 can provide developers some solid business advice.

Here's why the metaphor works so well. First, there's money to be made. A lot of money. Analysts at Canalys (*www.canalys.com*) estimate that revenue from app stores will top $36 billion in 2015. Similarly, analysts at In-Stat recently put out a market alert (*www.instat.com*) that mobile app download revenues would likely surpass $29 billion in 2015.

So much for the gold. What about the rush? Well, there are a lot of app developers, but no definite numbers. (Guesses range from 40,000 to 100,000 worldwide.) One thing is for sure: The number of developers getting in on the action is increasing at a rapid rate.

And, of course, developers aren't the only ones getting in on the gold rush. Advertisers, agencies, app stores, and technology vendors and brands are also panning in the mobile valley. Despite there being a finite amount of gold—with most "prospectors" only getting to see a few kernels—there is no end to the flood of independent developers setting their sights on the day they "strike it rich." Or, in other words, the day they replicate the success of Angry Birds, make their millions, and retire.

Golden Opportunity in Apps

In the California Gold Rush, billions of dollars (in today's money) worth of gold was recovered, but only a tiny handful of individual miners (the bedroom/garage coders of the app development world) made a substantial profit.

A slew of recent studies/developer polls show that individual developers aren't making much for their time during the current app gold rush either. In fact, the tech news site GigaOm (*www.gigaom.com*) has reported that publishers are making only $8,500 a year on average for their App Store efforts. It seems like an untenable situation for the individual miner, and in most cases it is, but the dream of striking that Angry Birds seam keeps them digging.

Developer Obstacles

Let's look for a moment at what's holding the individual app developer back during this gold rush period. The most obvious factor is their singularity, or oneness. Individual developers only harness their own earning power; successful business people understand that real success results from harnessing the talents (earning power) of many individuals.

Individual developers also tend to be extremely time-poor, which comes with a number of disadvantages. First, they don't have the capacity to take on the kind of app projects—often on behalf of big brands—that garner the best rewards. Such projects require fast cross-platform development (building an app for both iOS and Android mobile platforms). Most individual developers cannot resource these, even if they're able to stay up to speed with multiple programming languages.

Another issue for the bedroom/garage guys arises during the postbuild phase. Put simply, most developers lack the necessary marketing skills or contacts to make sure their apps are discovered. Instead, they publish their apps to the app stores and cross their fingers and hope their hard work finds a large enough audience.

So what can developers do to level the playing field? The answer is simple: band together and work together, just as many of the prospectors who flocked to California for the Gold Rush learned to do.

Assembly Model Innovation

The Chelsea Apps Factory (*www.chelsea-apps.com*), based in the United Kingdom, has an innovative model that provides a valuable blueprint for the mobile industry, and one you may want to consider. The idea is to bring together a community of developers, marketers, and investors under one roof, with the contribution of each helping to generate an app-centric creative atmosphere for all. Developers are welcome to be a part of this virtuous circle and are encouraged to take space in the factory (for as little or as long as they want) to work on their own projects or on projects managed by the Chelsea Apps Factory on behalf of its clients.

Since it opened its doors in March 2010, the Chelsea Apps Factory has gained serious traction in the community, matching developers with projects from a number of blue chip clients, including Vodafone, Telegraph, RBS, and CNBC. But it's not just about hands-on developing. There is also an emphasis on consultancy as a service in its own right (rather than as a form of lead generation). Essentially, this consultancy offers advice that includes perspectives from the whole community. In practice, ideas are paired with developers, and developers with fully formed apps are paired with marketers and enterprises. The successful approach takes apps development out of the garage and into the factory.

Sections of this chapter were contributed by: Matos Kapetanakis, Vision Mobile; Sam Chan, WIP; Linda Daichendt, MTAM; Dan Appelquist, BlueVia; and Mike Anderson, the Chelsea App Factory.

CHAPTER 5

What You Need to Know Before Building

If you ask any app developer how much it costs to develop an app, you'll probably receive a few different answers. There are many factors to take into consideration when developing a budget for building your mobile app. You need to decide your overall goals, what value you are looking to receive with this investment, then analyze the benefits you are seeking to achieve and determine the cost. There are also many technical requirements to take into consideration when embarking on hiring a person or a team to build your app. These requirements need to be thoroughly evaluated. If you have done sufficient planning, this journey will be very rewarding.

Value, Benefit, and Cost

From a business perspective, it is always better to focus on the value of the service and the benefit to your business than to focus purely on the cost. Most likely, you already understand the value of the service and how it can benefit your business, but there are always costs.

There are a number of reasons that app developers appear vague and noncommittal when asked how much it costs to develop an app. First, a large number of app developers feel they need to be very secretive about their prices. Second, it is difficult to provide a price without knowing the specific details of the app.

Type of Apps

The first factor in determining the cost of having an app developed is the type of app. So, what types of apps are there? Here is a list to help you figure out where your app fits.

- **Data-driven app:** This is an app where the data is dynamic. This means it's either stored in a local (on the device) database, or retrieves the data from an external source.
- **Games:** Have you heard of Angry Birds? Of course you have! That is a game, and there are other games, too. Games can range from the simple "thinking man's game" through to the very complex "action" games.
- **Device app:** A device app is an app that makes use of the hardware to provide its core functionality. That means it taps into a part of the device, such as the camera, accelerometer, or GPS.
- **Bespoke functionality app:** A bespoke functionality app is an app that is designed to provide a solution to a specific need or problem. It would be an app that may include data-driven features, device features, and bells and whistles, but it will also include very dedicated and specific features that will only be relevant to provide the solution to the defined problem. For example, if you wanted a time management app like a calendar, most of the features could be defined under the data-driven app type, but this app would also include very specific time engine functionality.

Design Costs

The overall production of an app is not just about the functionality that the app offers; the design of the user interface is also very important. A coder (a person who designs and writes and tests computer programs) and a graphic designer will both have different opinions regarding which one is more important, but some points stand out. Essentially, apps that don't have the functionality expected by the user or are not presented well, and therefore don't engage the user, are going to be unsuccessful. The design and the functionality of the app should be given equal importance.

How Much You Should Expect to Pay

This amount will depend on a number of factors. Chief among these is the scale of your project in the first place. Obviously, the bigger the scale of the project, the more graphic design work will need to be done.

Other Cost Indicators

It's not just size that matters. You need to be clear about:

1. **The number of device types and operating systems.** If your app is going to be available on more than one type of device and operating system, say, across Apple, Android, and Windows Phone, then there are different graphic requirements. While there is some obvious design overlap, keep in mind that the production of the separate graphics for each device type will increase the costs.
2. **The device types.** Even within a single manufacturer's device, there are different requirements between the individual devices. For example, the iPhone 4S and the iPhone 5 have different graphics requirements, and there may be differences between the iPhone, the iPad, and the iPod Touch because of differences in the app design. Similar distinctions exist on Android devices. Again, there is some obvious design overlap, but the production of the separate graphics for each device will increase the costs.

As a rough guide to the design costs, you should expect to pay between $450 for a small, simple app and $4,500 or more for a very large, complex, and graphic-centric app. The graphics for a game are in a different league entirely, and you could pay as much as $7,500. You should expect the costs to increase by 25 percent to 50 percent for each additional device type.

Development Costs

Again, many factors will determine the development costs of your app. The scale of the project will have an impact on the costs; a small app project that can be handled entirely by a single developer will be cheaper than a larger project that will require additional resources, such as project management, and a team of developers. As a rough guide, the details below show what you could expect to pay for each type of app:

- Simple app: $2,000 to $6,500
- Data-driven app: $6,500 to $45,000
- Games: $9,500 to $22,500
- Device app: $2,000 to $22,500
- Bespoke functionality app: $6,500 to $22,500

Total Costs

All of the costs associated with the production of your app will be covered within the design and development costs. The only other costs will be for creating the app store account ($99 for Apple apps, $25 for Android apps), and the costs associated with marketing your app.

Hourly Rate vs. Fixed Price

There is a tendency to attach an hourly rate to software development resources, which is fine when you are paying someone for their time as an employee, or psuedo-employee. However, in the context of getting an app development business to produce an app for you, you are paying for the skills, knowledge, experience, and professionalism of the business, and you are more likely to be given a fixed price for the project rather than an hourly rate.

Depending on the scale of the project, the fixed price for a project will encapsulate a number of separate resources, designers, developers, project managers, marketers—the list goes on. Each has a distinct role to play, committing a different number of hours to the project, and each earning a different amount internally.

On a smaller project, the same individual may actually perform many of these roles, but each role will be given a specific value and earning rate internally. This makes it very difficult to provide an overall hourly rate for a project. Based on the market rates for the skills utilized by app development businesses, you should expect to pay between $95 and $140 per hour for a good app development business to produce a very high-quality app for your business.

Bargain Price

One last warning: If something seems too good to be true, then it probably is. If you are given a price to produce your app that is significantly lower than expected (based on the previous information), then you need to ask why.

There are some legitimate reasons that an app development business may offer you a lower price than expected, such as interest in associating themselves with a specific industry sector, to build up knowledge and experience in an area of the technology that they have had limited exposure to, or even as a favor to a friend; however, they will be open about their reasons for offering a lower price.

If they are "just cheaper," they will have to be making up the difference in other ways, such as churning out a high volume of mediocre apps, which is never beneficial to the client/customer. Remember: focus on the value of the service and the benefit to your business, understand what you are getting for your money, and not just on the cost.

Risky Business

There are several areas around the business of app development that you need to be aware of. Otherwise, you might find that events quickly spiral out of your control. To avoid a regrettable—and expensive—app development disaster, you have to take steps to ensure your app, and your success, belongs to you and only you.

Ask any app developer to tell you about one of the most common and memorable conversations they have had with potential clients. No doubt they will share one that goes something like this:

[Potential Customer]: I have an idea for the next "killer app;" can you give me a quote please?

[App Developer]: Certainly, tell me about the idea.

[PC]: I can't, you'll steal it.

[AD]: I really won't. I need to know about the idea before I can tell you how much it is likely to cost.

[PC]: Can you just give a rough ballpark figure?

[AD]: Okay, it will probably cost less than a ballpark.

This is where the value of a nondisclosure agreement (NDA) presents itself.

ESSENTIAL

An NDA is a legal contract between two or more parties that details confidential material, knowledge, or information that the parties wish to share with one another for specific purposes but wish to restrict access to by third parties. It is a contract through which the parties agree not to disclose information covered by the agreement.

In the context of discussing your app idea, "killer" or otherwise, with an app developer, the NDA will give you peace of mind that the app developer is not going to steal your idea or pass any information you have discussed in confidence on to anyone else. Most developers have enough ideas of their own without needing yours, but the NDA makes this assertion official and professional, and more importantly, legal.

The app developer will usually have a template copy of an NDA you can use if you don't already have one of your own. However, as with all legal documents, it is always worth having the wording checked by a legal professional.

Ask Questions

What is likely to happen if you don't ask the developer to sign an NDA? In the vast majority of cases the app developer is not poised to steal your idea, nor is he or she likely to pass on any information that you've discussed. In other words, no need for you to panic.

What to do if the app developer *refuses* to sign an NDA? Well, this is where some bells should go off. You should start by asking for a good reason why the developer doesn't want to sign an NDA, but you should also be aware that there are some quite legitimate reasons why a developer might choose not to sign an NDA with you or anyone, period.

For example, it may be that the app developer wants to avoid the potentially uncomfortable situation that can occur when two great minds think alike. Think it through. If the developer signs the NDA and then realizes that your idea is very close to what he or she might already be working on, or planning to work on, then your NDA puts the app developer directly in the firing line for a future potential lawsuit.

Some app developers are happy to sign NDAs, some will refuse to sign an NDA for a discussion in the very early stages, and some will refuse outright. You will need to make a judgment based on the reasons your developer gives, and then decide if you want to pursue the app project or look for another developer.

Avoid Being Vague

After you have explained your app idea to the app developer, and the developer has made it clear he or she understands what you want and can produce it for you, it's your next task to make 100 percent sure the developer *really* understands what you want. A communication breakdown here can mean the difference between success and failure later in the app development project.

Don't assume that the app developer is on the same page with you from the get-go and understands each of your individual requirements. App development is a process that usually begins broad and grows deeper with each discussion or exchange. Therefore, the initial discussions, in particular, can start off vague. Don't let them!

Part of the ambiguity is due to the size and scope of an app project. There will no doubt be loads of information in your head around how to accompany the user on the journey through your app, how to encourage a specific call to action, and how to enable viral marketing and sharing. So it's natural that you leave out pertinent information along the way.

However, good app developers will not let you skip steps. Instead, they will pepper you with questions to draw this information out. If your app developer is not asking questions, either you are very good at getting your ideas across, or the developer is not really clear about your requirements. Make sure it's not the latter.

No app blueprint is carved in stone. Be flexible, and accept that your app idea starts out as just that: an idea. It is possible that you want your app to do what no app can, or you want it to function in a way that may not be the most efficient or best approach.

Again, good app developers will not be quiet. Instead, they will make suggestions for alternative functionality, or implementation approaches. If your app developer is not making suggestions along the way, then it could be a warning that the developer simply doesn't understand your requirements.

While the instant feedback you get during the initial discussions of your app can be enlightening, the real indicator that the developer understands your project will come from the quote.

A proper quote will tell you more than just "it will cost this much and take this long." It will reflect a deep analysis of your requirements. And it will provide a breakdown of your requirements, detailing what is required, the various options for implementing each of the requirements (or groups of related requirements), the app developer's recommended option, and the reasons for that recommendation.

In other words, a quote is not a price; it's more like a planning document that maps out what your app will do, how it can do this, and how the app developer will make sure it does. This paperwork is often referred to as the technical overview document. In it, app developers will include suggestions for alternative or additional functionality to make sure your app shines. All of this is aimed at taking your app idea from an "idea" to a formulated technical overview that allows any developer to read and understand your requirements.

Getting this level of feedback from the app developer should put your mind at ease. It is proof he or she understands your idea in the same way that you do. If you don't get this from your app developer, you should ask for it.

What Are the Payment Terms?

Be sure to speak openly, and clearly, about money. It is very unlikely that app developers will forget to tell you what their payment terms are, how they expect to be paid, and when they expect to be paid. In fact, most app developers will outline their payment terms as part of the technical overview documentation, or in the early stages of initial discussions. It's critical to have this discussion early on in the project to identify and talk through any issues that arise.

Standard Payment Terms

While there are no "standard" payment terms in the fledgling app development businesses, some common terms apply. The most common practice is for app developers to request half of the payment before the project is started and the remaining balance at the end of the project. This is common for shorter, smaller, and well-defined projects. For larger, more complex projects, however, the app developer is likely to suggest milestone payments, so that regular payments are made as areas of functionality are completed.

As with all commercial transactions, there is always room for negotiation. But it is also worthwhile remembering that the payment terms are there to serve as protection for both parties.

In addition to the traditional payment terms outlined above, there are also other methods that will allow you and the app developer to reach a commercially viable compensation agreement. However, there are also some pitfalls associated with these options that you need to know.

Partnership and Compensation

If you go down the partnership route, then be sure you have a partnership agreement in place that lays out, in plain language, the roles and responsibilities of each party, and the expected recompense from the partnership.

A partnership may force you to share assets you would rather control. Take a revenue share agreement that allows each of you to have a percentage of the app revenue. Granted, you don't have to pay the app developer cold cash. But you may have to give the app developer a majority stake.

Why? Think of what each of you is bringing to the table, and a 60/40 split makes business sense. The idea for the app is yours, but the app developer in the partnership is taking on a significantly higher risk by going into a partnership with no guaranteed results and performing the development role on the off-chance that there will be a significant return on the investment. Therefore, don't be surprised if your potential partner wants a higher percentage of the return than you were expecting.

Your partner may also want a larger say in how the app is developed and marketed. Under traditional payment terms the app developer offers advice and guidance based on their knowledge and experience, but it's *your* decision to take the app in the direction that you choose. This is a completely different story when you take on a partner, especially when that partner is also the one developing the app. You may find that your artistic control is diminished, and, in the worst case, the app developer could ignore your opinion completely.

Share the Wealth

It is surprising how often the discussion drifts to approaches around paying the app development costs out of the app sales. The payment option may look good on paper, but don't expect the app developer to share your vision or enthusiasm. More often than not, suggesting that the developer take his or her fee from the future app sales will be met with laughter or incredulity. Sometimes, profanity may be involved too.

But, before you judge the app developer, think about what you are really asking. This is how your suggestion translates in the mind of the app developer:

"I have a rough idea for an app, but I don't want to take any of the risk. I would like you to work, using your skills and knowledge, on the project and, if it is successful, you will be paid bit by bit as the app sales come in, until your normal fee is paid. Then I will reap any future rewards without having had to take any of the risk. However, if the app doesn't make any sales, then only you will have lost out. So, either way, all is good for me."

Clearly, this option can get your business relationship off on the wrong foot. So, think carefully before you suggest it.

No matter which payment terms you choose, make sure that there is no ambiguity. This means ensuring that the payment terms and any allowable variations are clearly and precisely defined and included in the contract.

Owning What Matters Most

You're likely—and advised—to spend a lot of time working out the proper payment terms. But there's an even bigger issue that requires your attention: source code. Put simply, source code is all the programming language and commands that power your app. It's common to assume that paying for the app to be developed automatically entitles you to own the final source code. But this is not necessarily the case.

It's a thorny topic, and it's important to be clear about the real issues and scenarios that can arise. One concerns whether the developer should hand over the physical source code that has been written during the development of your app. The other is around the copyright/intellectual property (IP) rights of the source code and elements of the app, such as graphics, that can have IP rights assigned and/or transferred.

Who owns what is not always clear cut. A rule of thumb: The developer of the software automatically retains the IP rights, unless those rights are assigned to a third party. As with any and all legal considerations, you are well advised to seek professional advice.

This is also why you should lay this out in the contract before the app developer starts the work. As long as you ensure that the IP rights are assigned to you in the contract, then there should be no future wrangling over the ownership of the source code and other app elements.

Source Code Is Essential

So, if you do the right thing, you will have a contract in place that states you are the legal holder of the IP rights. But does this also mean you are the owner of the physical source code? Well, yes. And that's good news because you will almost certainly need the source code in the future.

In an ideal world you will have found a fantastic app developer whom you want to employ for all future updates and/or the work of developing additional apps.

But that relationship, like wedded bliss, is rare. If the relationship sours and you decide to break up with your app developer, then you'll want to leave with the source code in hand. This is the only way you will be able to pass the source code on to another developer and pick up where the first app developer left off. Leave without the source code, and you will need to start the entire app development process all over again with the new developer just to get back to where you were with the app developer you let go.

The battle over the source code can be like a messy divorce—and worse. Couples may fight over the car and kids, and app developers may insist on keeping the source code. (This is why it is imperative to lay this out in a contract before you start your app.)

Why do app developers want the source code in the first place? In a word: power. If they own the source code, then they can effectively force you into an ongoing relationship. This means you must go to them to make any future changes or updates at whatever price they want to charge you. In addition, the app developer may want to keep the source code as a blueprint so they can recreate the app again and again for other clients. No matter the motivation, it's bad news for you, your app, and your business.

Have You Got Exclusivity?

This goes hand-in-hand with the previous section. If you don't own the source code, do you have exclusivity? Even if you do own the source code, do you automatically have exclusivity?

To start: What is exclusivity? Exclusivity ensures that the developer will not sell the source code to someone else to create a competing product, nor will the developer re-create your app as a competing product directly for another client, or for him- or herself.

Even if you own the source code, you should still ensure that you also have exclusivity. But be aware that exclusivity will only ever apply to your specific implementation of your app idea. It will not cover variations of it.

Do You Really Own the App?

This is not a dumb question. If your app developer puts his name on the app, it belongs to him or her, not you. Specifically, the app belongs to the legal entity, be that an individual or a business, that has signed the developer agreement for the developer account that hosts the app.

Put another way, if the app is on the app developer's account, it's his or hers—period. No matter the legal documentation or guarantees that you may have in place, the rules of all app stores are clear. If you ever change developers and want control of your app, it will have to be removed from the app store altogether and resubmitted under your own developer account.

ESSENTIAL

You will not be able to resubmit the app using the same branding, so if you had a great name for the app, you'll have to change it even if the app uses exactly the same source code. In the eyes of the app store, your app is a completely different app. Your app also has no history or track record, so any reviews or ratings that were associated with the app in the app store will be gone.

This also means that the customer base you have built up—fans who have downloaded the original version of your app—will not benefit from updates to the new version of your app. To be up-to-date with your app they will need to download the new app. If yours was a paid app, then these customers will also need to pay for it again. You can imagine that not many of your customers would be willing to literally pay for your mistake, so take steps to avoid it at all costs.

Adjusting to Change

Your app works by communicating with the underlying operating system through APIs. In other words, your app depends on the decisions and directions of the operating system owner. Of course, you can't know ahead of time how Apple, BlackBerry RIM, or the Android community will change, so there are also no guarantees that your app will still function.

Fortunately, the changes won't come out of the blue. Your app developer will know about any potential changes and whether they will impact your app before the system updates are released.

Handling these changes will often form part of the maintenance agreement that you have with the developer, but you should clarify this when you negotiate your maintenance agreement. If these updates are not included in your maintenance agreement, then you will need to come to a separate agreement with your app developer, laying out how these issues will be dealt with and the costs involved.

If your app has been developed using an app generator tool, a tool that creates a mobile-ready application, then don't count on a quick fix. This is because an app generator tool makes use of third-party libraries, and depends on these third parties to act fast when the underlying operating system is updated. Your app developer will have to wait until the libraries are updated first. After the libraries are updated, the app developer can update your app. This can take time and potentially leave you with an app on the app store that doesn't work.

Contributed to by Jez Harper, Tús Nua Designs; Alred De Rose, Tego Interactive; Paolo De Santis, Chupamobile; and Magnus Jern, Golden Gekko.

CHAPTER 6

Developing Your App

When it comes time to think about the actual production of your app, you will no doubt still have some questions: Who will you get to develop the app? Should you develop the app in-house? If you outsource the project, how do you find a good developer? Are there any ways to reduce the overall cost? You will find the answers, or at least the information to guide you to the answers, in the following sections.

Who Will Make Your App?

This is not a simple choice of in-house or outsource. It's not something that should be decided on the flip of a coin, it should be given very careful consideration. Even within the "in-house" or "outsource" choices there are a number of options. For in-house production, will you do all of the work yourself, or use your existing staff, or bring in short-term contractor(s) to do the work? And for outsourcing, will you utilize offshore development services, find a freelance app developer, or make use of a mobile app development agency?

Those are not the only options either. Don't forget that an app is made up of many parts and there are many different skills involved in the development of an app, including project management, app architecture design, UX/UI design, coding, and testing. Any combination of these can be performed separately, either in-house or as an outsourced service.

Should You Go In-House?

This is definitely an option that should be considered alongside the other options. Whether it is a suitable choice will depend on a number of factors, such as budget constraints, availability of the necessary skills, the type of app being developed, and time constraints. It is important to understand that none of these factors should be the sole consideration in making the decision.

Budget Constraints

This is the common factor used in determining what choices people make in their everyday business lives, and unfortunately, it is all too often the only factor used to make decisions, and that can result in some undesired outcomes.

There are costs involved whether you outsource the entire project, parts of it, or keep the project in-house, even though the costs may be less obvious when you aren't handing over finances to an external third party. Producing the app in-house is not necessarily any cheaper than outsourcing the project, for a number of reasons:

- **Reduction in normal productivity:** Assuming that neither you nor your staff sit around with nothing to do, whoever is involved in the app project will not be performing their normal duties, or at the very least will not be performing them to the same extent that they would be normally.
- **Time to market:** The reduction in normal productivity will be compounded by the fact that it almost certainly will take you longer to complete the app project than many of the outsourced professional app development options.
- **Cost of failure:** While you may appear to be saving money by keeping the app project in-house, if you do not have the necessary skills and experience to produce the app to the same level of quality that is produced by professional app developers, then your project is less likely to succeed and the savings that you have made will be regarded as a loss.

There are ways of reducing the costs without reducing the quality of the output, and you should certainly consider this before deciding that you don't have the budget to outsource the project.

Skills

There are two questions to ask regarding the skill set already available in-house. First, the obvious question is: Do you already have the skill set available to produce the app? An often overlooked question is: Is it cost effective to develop the in-house skill set so that apps can be produced in-house?

An answer to the first question will depend on the type of app that you are planning on producing and the in-house skills you have. For example, if you are planning on creating a mobile-enabled website (web app), or even a hybrid app using a tool like Appcelerator's Titanium (*www.appcelerator .com*), and you have competent web developers in-house, then you possibly have the necessary skills at hand.

However, you will need to consider more than just coding ability when judging the available skills. There is more to producing an app than just coding. Designing for the user experience (UX) and designing the user interface (UI) require a different mindset when creating apps for mobile devices, as the way they are expected to behave and be used is different from how normal desktop software or a website would be used.

These are, of course, relatively small changes required by the thinking of the designers and developers involved in the app project, and these may be areas where it is worth considering up-skilling the existing in-house skill set. You may even want to consider training existing developers to a level where they can create native apps, but that option will likely be reserved for businesses that are planning on creating multiple apps, or feature-rich and responsive apps.

Time Constraints

Do you have the time to commit to producing the app in-house, especially considering that it will more often than not be quicker to have it produced by a professional app developer? Do you need to get the app out to market quicker than your in-house production will allow?

Developers are, in general, optimistic. Whatever your developer gives you as an estimate for how long something will take, double it. This is common practice with project managers.

The app production process usually breaks down as such:

- Predevelopment activities, or scoping: two to three weeks on average
- Development: six to ten weeks on average
- QA: two weeks after development finishes
- App distribution and deployment: one week

Factoring Everything In

Everything previously mentioned may seem to suggest that the best option is to avoid developing your app in-house, but that doesn't mean an in-house project isn't a viable choice, you just need to ensure that you are considering all of the factors when making your choice.

The Cost of Development Methods and Models

You get what you pay for with apps, and contingencies need to be allowed for to make sure there is a buffer, as decisions sometimes need to change or as a new (business) requirement may need to be added during the development phases. Before you look for a developer, you need to be clear about your options and requirements. It may be that off-the-shelf tools will do the trick. Or it may be that you need to work with an agency.

Here are six paths you can take:

Do It Yourself (DIY) App Development Tools

DIY development allows anyone to develop a mobile app based on templates without the need of development skills. The advantage of this is that you can launch an app at a very low cost; however, the downside is that you are very restricted by the templates and functionality and you still need to have a pretty good understanding of apps and or great designers and creative resources to develop and launch a great app.

Examples: Mobile Roadie, Swebapps, Kanchoo, and AppBreeder

Cost: from $500 for the tool, plus your own time

Offshore Development

Most big corporations go offshore for some of their development and operations today, whether it is direct or indirect through a consultant such as IBM, Accenture, or Sapient, for example. The advantage for well-specified information technology (IT) offshore development and production development to India and China can be a big cost saving. However, the disadvantages include longer lead times, greater overheads in communication and specification work, and less proactive solutions that don't address your local market. It also takes time to learn how to manage an offshore development efficiently. If the price seems too good to be true, then it probably is.

Examples: There are 100+ developers in this category.

Cost: $15–$30 per hour

Freelance Developers

Using contractors or freelancers is popular among midsize companies, as they have the freedom to bring in specialists for a limited period of time without taking the risk of a permanent hire. In addition to this, it is usually easier to check the references of a freelancer when hiring. The disadvantages are that projects can take a lot longer and cost more than expected, the person may disappear with all the knowledge at any time, there could be a single point of failure if something goes wrong, it is limited to the person's own expertise and experience, and lastly it is difficult to verify the quality of the freelancer's work and ensure a quality testing of the app.

Examples: There are thousands of freelance mobile developers in the United Kingdom and United States.

Cost: $40–$50 per hour

In-House Development

Big companies such as Amazon, eBay, Google, and Walmart hire, train, and frequently acquire their own developers, since the mobile app platform is considered to be a core part of their business. The advantages of in-house development includes keeping knowledge in-house, less time spent on contract negotiations, and potentially slightly lower development cost than when using third parties. The disadvantages are that recruiting and retaining skilled developers is very difficult; salary costs are high or very high in areas such as Silicon Valley, New York, or London; less experience and input from other businesses and projects; and frequent major project overruns, as there is less accountability in internal IT departments.

Cost: Salary ranges from $40,000 to $200,000 depending on location (United States and Europe)

Mobile App Development Agencies

There are hundreds of mobile app development agencies, ranging from companies with 2 people to those with more than 100 people, and most are made up of people with a web development background. The advantage of these specialized agencies is that they will usually commit to a fixed price and timeline for projects, which reduces risks for the customer. The disadvantages are that your expertise and quality varies a lot between agencies, development skills for more complex tasks such as back-end integration are limited, and costs frequently rise during the project due to unplanned changes and issues.

Examples: The Icon Factory, Chelsea Apps, Rancon Apps, Sprpd, and Idealapps

Cost: $50–$150 per hour

Premium Agencies

There are a few bigger agencies that have a reputation of consistently delivering mobile services and marketing campaigns with great user experience, innovation, and creativity. These agencies manage complex back-end integration and projects across multiple operating systems and territories and launch industry-leading mobile services. These agencies consist of a mix of experienced digital agencies that have gone mobile and mobile specialists. The advantages are that they have processes, people, technology, and quality assurance based on years of experience, and their reputation doesn't allow failure. The disadvantage is that hourly rates and minimum budget commitment will be higher.

Examples: AKQA, LBi, Razorfish, Ustwo, and Golden Gekko

Cost: $100–$250 per hour

ESSENTIAL

Depending on your app, it might be worth considering working with a partner that has built an app of this kind before. Code reuse can save some costs. But it can also be difficult, so there is no 100 percent guarantee that it will reduce cost or timelines.

Factors That Impact Cost

Once you've chosen what kind of development partner/method you want there are still four key factors that will impact the cost:

1. Back-end integration complexity including maturity and documentation of APIs
2. Functionality in terms of number of features, complexity (e.g., augmented reality) and whether it has been done before
3. User experience requirements by client ranging from simple to very detailed wireframes, advanced interaction design, and pixel-perfect design, which requires a lot of optimization
4. Quality expectations in terms of performance, reliability, error management, and target devices to be fully supported

Unfortunately, there is no simple answer to the question, "What is the cost of developing a mobile app?" The cost can range from $1,000 for a DIY app to $41 million for a mobile service that is the basis for a mass-market consumer business.

ESSENTIAL

There are a few reasons that an app development business may charge less than market rates for developing an app. But don't jump at a bargain. It could be that the app maker has no long term plans to remain in the industry (or support *you*) and is only trying to make a "quick buck" by charging a lower rate so that they can secure as many jobs in as short a time as possible and then close shop—forever.

When choosing the company or developer you want to work with, it's important to check their references and ensure they have the experience and the aptitude they claim. Ask for several references, contact them, and make sure what they've worked on is relevant for you and that the developer or team you are getting actually carried out the work. Check what part of the work they carried out, what their responsiveness was like, and their delivery compared to timelines and budget. This will increase your chance of success a great deal.

Stick to the Budget

Whether it's an internal or externally developed project, you should agree on a budget for time and resources (with some buffer for changes) and stick to this. It's always better to get to a point when you deliver what's available at the time, launch or do an internal pilot with this, and close the project. Then agree on a second phase with a new budget and timelines. Otherwise app development projects have a tendency to go on forever, as there are always things that can be improved.

If you are working with an external developer he or she will most likely manage changes to the scope and timeline through formal change requests, which will incur additional costs. Plan for this from the beginning.

How to Find a Good Developer

If you've decided that to build the app you want the tools can't come from the shelf, the next step is to find an app development business to produce your app for you. There are a lot of people who can develop apps, from the school kid down the road to the businesses who "also do apps" to the dedicated app development businesses.

ALERT

The app development market currently favors the app developers; there are lots of businesses wanting to have apps developed and not many good app development businesses to develop the apps. Unfortunately, like many industries that experience growth, there will always be businesses that see the opportunity to make "easy money" rather than provide a valuable and reputable service.

So, how do you find the diamond in the rough? There are a number of things to look for in a good developer.

App Development Is Their Core Business

So why should you choose a business that develops apps as their core business? There is a big difference between being able to do something and being able to do it well. It is an industry that is always changing, and a significant amount of time needs to be dedicated to staying up-to-date with changes to the development tools and software development kits (SDKs).

A business that develops apps as their core business will be dedicated to ensuring that they stay up-to-date so that they can offer the very best service.

Focus on Solutions

You've probably heard of people who get tattoos that contain misspelled words. Why don't the tattoo artists tell the customer before creating a permanent salute to illiteracy? It's a similar situation with app developers.

There are some app developers that will create exactly what you ask for, regardless of whether it is the most appropriate solution to meet your needs. This could either be because of a lack of knowledge or experience on the part of the app developer, or because they are only interested in the one job from you and not focused on providing a service.

One thing is certain: If you are not getting detailed feedback from the app developer, even at the very early stages of the app development, you will be paying for coding, and not for the benefit of the app developer's knowledge and experience.

Proven Track Record

It is very easy to "talk the talk," especially when discussing a topic with someone who does not have in-depth knowledge. It is far more difficult to "walk the walk," so it is very important that the app developer can prove that he or she is capable of producing the very best results when developing your app.

There are a number of ways for app developers to prove their capability. They can show that they have developed apps already, either for themselves or for others, but preferably for others. They can provide testimonials from previous clients/customers, or even better, they can put you in touch with them directly. Finally, recommendations from other businesses that have had apps developed can be a very useful indication of the quality of an app development business.

Beware of Tool Users

There are short cuts both in business and in your personal life, and very rarely do these short cuts produce good results. This applies to the app development industry as much as any other; there are tools that can be used to produce very simple apps without having to write a single piece of code.

Because of this, there are people who will call themselves app developers when, in fact, they aren't. There is more to any software development than the end result. A software developer (as opposed to a tool user) will be able to fully understand your requirements and any potential issues or implications from implementing your requirements, will be able to make alternative suggestions and explain a more appropriate solution, will understand how to resolve any potential issues, and will be able to provide exactly what you want in your app.

Rise of the Mobile App Component Marketplace

Many app developers are choosing to make their work easier, and potentially more profitable, by making use of those high-quality, prebuilt, and vendor-supported app components offered via mobile app component marketplaces. Verious, for example, enables app developers to license prebuilt, pretested libraries to accelerate time-to-market, access third party content, and increase in-app monetization.

Appcelerator, by virtue of its Titanium platform and its large developer network, is also a major player in the space. The Appcelerator marketplace is stocked with products including gaming modules; solutions for integrating PayPal, DropBox and Millennial Media; professionally designed app templates; UI controls; and much more. Publishers are adding new products to the marketplace every day.

At the other end of the spectrum, Chupamobile (*www.chupamobile .com*), a community marketplace where app developers can sell, discover, and purchase source code packages and components for mobile app development, is also gaining significant traction. Chupamobile aggregates and categorizes quality code and UI elements for a variety of mobile development frameworks, from the widespread native iOS and Android to the well-known cross functional jQuery Mobile, Sencha touch, Corona, Phonegap, and Titanium.

According to a survey of app developers conducted by Chupamobile in June 2012, over 50 percent of the respondents reported saving at least 30 percent of their time by using and integrating external components. In addition, mobile app components that normally need 30–40 building hours can be purchased with less than $20, and can save up to 90 percent of development time in some cases.

Clear Communications

A good app development business will explain the entire process of creating an app before it has even started and will guide you through each step, providing clear instructions for any part of the process that needs to be completed by the client/customer.

It is easy when dealing with an outside party to lose sight of what is happening and to feel like you are not in control. A good app developer will understand this and will ensure that you are kept informed completely during the process of developing your app. The people working on your app should also be open to communication from you and never make you feel like you are inconveniencing them.

ALERT

A good app development business will correct any issues that have arisen due to their code, at their expense. They will also take part of the responsibility for any issues that have arisen due to how the initial specification of the app was developed, because a good app developer will have analyzed the initial specification and should have spotted any potential issues and advised accordingly.

Depending on the scale of the issue, the costs are likely to be split between the app development business and the client/customer, although if the issue is a small one, the developer will probably just fix it at their cost.

Any issues that arise because of changes to the underlying device operating system are not issues that can easily be foreseen, and you should not expect the app development business to take responsibility for the costs of correcting the issue. However, any good app development business will expect these issues; in fact, they will more than likely be aware of the issues before you, or the users of your app, are. They will inform you of potential issues before they arise, so that appropriate action can be taken.

Sections of this chapter were contributed by Alfred De Rose, Tego Interactive; Jez Harper, Tús Nua Designs; Magnus Gern, Golden Gekko.

CHAPTER 7

Designing Your App

There are no hard-and-fast rules for designing a good app. On paper it all appears to be quite simple, provided you follow some basic steps, but sticking to the basics is not always the way to deliver on the promise of good design. Truly good design principles and attention to detail will result in a compelling app, and one that your customers will interact and engage with time and time again. Making the right decisions in this critical phase of your app journey will play a huge role in encouraging brand love from your consumers.

Starting Your Development Career

Many app developers start this process with a singular focus on app functionality. While this is important, try not to leave the design of the user interface (UI) as an afterthought. It's an understandable oversight. Overlooking UI in this phase can mean the difference between a good user experience and a really bad one.

Designing for Simplicity

Regardless of how complex your app is going to be, whether it is going to be crammed full of fantastic features or a single-function app, it needs to appear to be simple. The word *appear* is key here. The app itself can be as complex as is necessary to provide the overall functionality the user will need. But it needs to look simple, clean, and easy to use.

Is it really important to design for simplicity? If you've never downloaded an app and thought, "I don't even know where to start with this," or where your first thought is "I hope there is a manual," then you have been very lucky indeed!

Unfortunately, many apps out there are not very straightforward. You need to do all you can to deliver to your customers an app that they can understand and use from the get-go. This isn't just about making the customer happy, which should be at the top of your goal list anyway. This is also about damage control.

People tend to form instant opinions of apps, and that process begins the moment they download the app. Once people have a negative opinion of your app, it's a mammoth task to change their mind. You will have to work very hard to win back the users that you've lost in those first few moments, so it makes far more sense to win them over the first time they use your app.

ESSENTIAL

Don't try to design the UI until the functionality and flow are clearly defined. If you try to get ahead, you'll just waste time redoing the UI design again and again.

Design Dos and Don'ts

How do you keep your UI interface simple? Here are some tips to guide you:

- Avoid cluttering up the view. Remember that while your app may have lots of functionality, your users are unlikely to want to use it all at once. Split your functionality into independent groups.
- Guide the user through a flow. If the user is expected to perform actions in a certain order, present them with only the necessary screen elements they need to be able to perform the current action. Once they complete it, and only then, present them with the next view.
- Avoid the use of text where possible, especially long wordy paragraphs of text. Anywhere you can use an image as a replacement for text—and maintain the meaning—you should do exactly that. This will keep the view clean and will aid in any localization efforts. However, this doesn't apply to any info pages you include in your app. If the user has gone to the info page, then they probably want to read about your app. Give them what they expect.
- Just because you can, doesn't mean you should. There are lots of "cool" UI features that you can add to your app: swiping entire pages up or down; left or right; swipe to the side to reveal a menu, or options; even 3D rotating table cells that you can swipe in any direction. These are very clever, and individually, very cool, but without careful consideration, your app will very quickly become an overwhelming cacophony of bells and whistles that confuse the user, hide the core functionality, and render your app almost unusable.

Be Consistent

Your app must be consistent. This means it needs to have a consistent look and feel. In practice, all the views, screens, and pages within your app should all look and behave in a similar fashion. This is not suggesting that a view containing a map should look the same as a view containing a list of items, but it does mean that you must make an effort to provide the user the same tools and views when the task is also similar.

For example, if you ask the user to select a date from a picker control in one part of your app, then anywhere else in your app that you require the user to select a date, you should present them with the same picker control.

Advantages to Similarities

In addition to maintaining a consistent look and feel throughout your app, it is also very important to compare the consistency of the look and feel of your app with other apps on the marketplace. In the world of apps, as with most software, familiarity is an asset. You want a user to be able to download your app and start using it straight away. They can do that instantly, and without having to endure any significant learning curve, because it looks and behaves like many of the other apps that they've already used.

Sure, your functionality and look should (and must) be different to meet the requirements of your customers, but the basics should be the same. A button should look like a button and do what a button does; and a navigation bar should look like a navigation bar and behave like a navigation bar. If you create your own button, it doesn't need to look like a carbon copy of the standard button, or anyone else's button for that matter, but it does need to be obvious to the user that it is a button.

In many cases, you will have to use your own judgment. But be careful not to reinvent the wheel. You should also avoid overdesign. Overdesign is where "reinventing the wheel" goes to a whole new level, where nothing, or very little, of the design bears any resemblance to anything in any other app. While this can look very good and very appealing while you are designing your app, the implementation may result in anything but a positive user experience.

Consider Context and Usage

Factor in how and where your app may be used. This is not just about location. This is about what your customer is doing when they access and use your app. Will the customer be seated or standing? Will they be static or moving? Each of these scenarios puts the pressure on you to design an app that goes with the flow.

Will your app be used for "snacking," or consuming a small amount of information at a time? If the answer is yes, then the UI needs to be designed to allow users to quickly open the app, get the data they want, and exit just as quickly. Will the app be used to immerse users in content or experiences? If the answer is yes, then you need to make it possible, and pleasurable, for the user to spend more time within the app.

Speed Matters

There are probably very few app users who have said "I wish this app was slower," but you can be sure that there are many who wished the app they were using were faster. This is a fast-moving world where there is an expectation that information is available at the press of a button and that it is available immediately. This is no different with apps. In fact, the expectation is that apps should be ready to serve the user with what they need, almost before they've even asked for it.

On the whole, certainly with native apps, this performance requirement imposed by the users is not a problem. However, there are a few circumstances where performance can be degraded by the app design. The first is where the app hasn't been designed for simplicity and too much information is being displayed at once, or the app is attempting to do too much at once. The second circumstance is when data is being loaded into the app from a remote source. Again, this will only be a significant issue if the UI design has not considered the time taken to load remote data and attempts to load too much at anyone time.

You can improve the perceived performance of your app by managing the amount of data it loads. Rather than loading a large amount of data and making the user wait until it is loaded, just load smaller chunks of the data as and when needed. This will reduce the perceived waiting time and giving the user the appearance that everything is happening *almost* immediately.

For example (and these numbers are simplified to explain the idea), if the app loads 1,000 items of data to be displayed in a table, taking a total of eight seconds, that would be an unacceptable time to wait for an app to respond.

However, if the app loads fifty rows at a time, and when the user reaches the bottom of the table, the next chunk gets loaded, taking half a second for each fifty rows, it would actually take two seconds longer to retrieve all of the data, but as far as the user is concerned, it is considerably faster.

Multiple Platforms and Devices

Whether you are developing native apps using the development tools and languages provided by the various operating system and platform owners, or you're developing a hybrid app using third-party libraries and app generator tools, you need to give careful thought to how your app will look and function on *all* of the platforms you intend to target.

And you need to address this early in the app process unless you want to waste, time, money, and effort later on. If you make the mistake of designing the UI for only one platform, you'll have to adapt it for the other platforms. Otherwise, it will look like it has been shoehorned in to fit.

Platform Designs

Each platform has its own UI guidelines that you are expected to follow. Unlike a developer with a desk full of different devices as test equipment, users tend to be familiar with a single platform and expect the apps they use to follow the same guidelines as the other apps they use. If you produce an iPhone UI design and then make an Android UI look the same, it will most likely not be adhering to the Android UI guidelines and will certainly be an unfamiliar UI to the Android users.

ESSENTIAL

Android is currently the reigning platform across all age groups. Android devices have varying screen sizes and resolutions that you need to contend with. With iOS you have to build to the highest retina resolution for the iPad3 and then halve it for all nonretina devices (iPhone).

Adapting to Tablets

More often than not, when you are designing your app, the customer expectation is that your app will be available on multiple devices, including tablets. You will need to factor this into your design consideration, and your overall business plan.

You have a few options to choose from. One, you can do nothing. Platform providers like Apple and Android are well aware of the advance of tablets, and will automatically stretch a phone app so that it fits the larger displays of the tablets. This is a viable option, but it isn't necessarily a good option, because it sends a clear message to the customer that you haven't put much effort into the UI design of your app.

Two, you can make the effort and design the UI from the ground up for an app that is intended to be used on tablets as well as phones. It's important to remember that the UI for each device—phone and tablet—should be considered as entirely separate interfaces and created accordingly.

With tablet form factors there is much more screen real estate to be utilized, which will have a significant impact on the design decisions you make. But it's not just about screen size. There are also significant differences in context and the way people use their tablets to access and enjoy content and services. One thing for sure: Bigger screens, coupled with the fact that tablet devices tend to be used with two hands, means you can borrow a lot of concepts from desktop applications to deliver a more immersive experience.

ALERT

Research is just beginning to reveal and detail how people increasingly use tablets as part of their daily routine. It's a market that is a work in progress, so stay abreast of developments by reading surveys and watching the trends.

If your target users are smartphone owners rather than tablet users, you may find it more cost effective to design and develop your app for the smaller form factor and leave the larger form factor app UI design and development to a later date.

Catering to Global Capabilities

Although it's important to design an app to make the most of the sophisticated features and functionality common to the newest devices, you should not ignore the requirements of customers with lower-capability devices. In many cases, it's these customers who will also reward your efforts to ensure the app will work well on the devices with repeated use and lasting loyalty.

Considerations around UI Design

While this should be addressed when you are in the beginning phases of the app process, there are still considerations around UI design. For example, say your app allows the user to take a photo to be used within the app. That is fine, but many devices, including the third-generation iPod touch, don't have a camera. If you are targeting these devices with your app, then the UI design should include and display an option that allows the user to choose an image from their photo gallery on their device, or at least select whether they want to use the camera or select from the photo gallery.

In other words, don't assume that everyone has the latest devices, chock-full with features and functionality, and design your app to appeal to the widest possible audience. If you were a department store, you wouldn't want to bar older-model cars from parking at your place of business. So why limit the number of customers who can download and use your app? Unless you know you only want to target early adopters with the newest, coolest devices, build apps that cater to a broad audience of users and devices.

Local Markets for Global Reach

Design for localization early in the process. People all over the world can discover and download your app, so build localization into your app and take steps now that will allow you to accommodate date formats, currency display, and localized images, without having to go back to the drawing board.

Consider that the word for "speed limit" in German is *Geschwindigkeits-begrenzung*. You don't need to be a rocket scientist to figure out that there will be many instances when the localized word won't fit in the space you designed for that purpose. What should you do? One of the best ways to reduce the issues created through localization is to simply minimize the use of text as much as possible. In a global market, a picture (in this case, an image), does indeed speak a thousand words.

For example, if there were two text boxes on a login screen, and one had a picture of a person next to it and the other a picture of a key, it would be fairly obvious that the user was being prompted for their username and password. No need here to worry about the pitfalls of translation.

Most of the time, simply creating a label that will wrap words and remain centered within the label, as well as the label remaining positioned correctly with any associated UI element, will be enough to handle most of the localization issues. However, for some words, this approach won't be enough. You will need to work closely with the person responsible for the localization translations to find an alternative word that will fit in the space available, while maintaining the contextual relevance of the word. Or you will simply have to change the UI design so that the longest localized version of the text will fit.

Words and More

Language is just one part of localization. You may also need to localize other elements of your app including:

- Display of dates and times
- Currencies
- Metric or imperial measurements
- Miles or kilometers
- Decimals or commas for the number separators

ALERT

Do not use machine translation services. They may work very well for a paragraph of text in an e-mail, but they do not work for apps and don't understand context. Instead, use the services of a professional app translation service.

Finally, do your research before implementing a design for every possible language. You will likely find that certain nationalities actually prefer to use the app in English. A business app is often just fine in English, but a utility app may need to be in the local language.

Visualize Your App

There are undoubtedly some fantastic apps that were completely designed on the back of a napkin. A good place to start designing your app is paper prototyping. You may feel that this is not a necessary step to define and map the user journey through your app, but it's a great way to be absolutely certain you have every step of the path covered. Crack this and you are halfway home.

Start Simple

Map the app journey out on a piece of paper; ask people you know (or don't know) if they would interact with your app in the way you are describing. Ensure that these core objectives are as simplified as possible and executed really well, because if they are overly complicated, people will simply not engage with your app this way in real life.

When you have a feel for how people will really interact with your app, try mapping out the app in more depth. Look over what you want to empower your customers to do and then match them with the unique features of the devices. For example, if you want people to find your store, you can tap into the GPS feature in the phone. If you want people to let their friends know that they have visited your business, you can pull their social feeds into the app.

Adding Flow

You can draw from your paper prototyping to lay the groundwork for your app. Start by using Post-it notes and scraps of paper to work out the flow of screens and how they all link together. Once you have the structure, you can ask your app developer to create wirefames, or screen blueprints. If you want to make your own wireframes, then you will need graphics software. Packages you can use are Balsamiq (*www.balsamiq.com*) or Omni-Graffle (*www.omnigroup.com/products/omnigraffle*).

Once you have the visuals covered, you can easily demo them on a handset. All you have to do is create a Keynote presentation with slides that are the same dimensions as the handset you wish to test on (for example, 640 x 960 for high-resolution portrait visuals), then copy and paste all of your screens into the Keynote presentation.

ESSENTIAL

Simplicity is key. There are far too many apps on the various stores that are overly complicated, confusing, and only ever live on people's desktops, never opened. Provide your consumers with a useful, engaging experience and they will love you for it.

You can then go one step further and put invisible shapes onto the buttons and clickable elements for each of your screens, and insert hyperlinks to these shapes that then link them to the correct screens in the presentation. When you view this on a device either through Keynote or a standard PDF reader, you will be able to link back and forth through your screen designs.

Think and Design Human

You can avoid a slew of common design mistakes if you put yourself in your customer's shoes, and then take a long walk. Always think of how your customer will use the app, and be aware that this may not match how you expect them to interact with your app.

Keep the user informed. Whenever the app is performing any task in the background, the app may appear to the user to be doing nothing at all. The user might then think "Crash!" Provide visual feedback so the user knows that something is happening. If the user is going to be kept waiting too long, simplify the process.

Don't lose sight of the overall aim of the app. It's very easy to get so carried away with making the perfect UI that you inadvertently shortchange the user by removing core functionality from the app, or reducing the functionality available to the user altogether.

User Experience

User experience (UX) is how a person "feels" about using a product, service, or system. These "feelings" can be categorized as either good feelings or bad feelings with the feelings generally determining whether the user experience is a good one or a bad one. Think of user experience as the perception left in your mind after a series of interactions or "dialogues" with services and products, through people, devices, and events; and any combination of those five.

Why Is User Experience Important?

Communication is shifting from one-way, that is, brands broadcasting to consumers, to multiway, brands and consumers starting conversations on several channels. Many businesses jump on the mobile strategies bandwagon by being present on trending channels like Twitter and Facebook but not actually having authentic behaviors that offer meaningful experiences; this underestimation of user experience hurts small- to medium-sized businesses more than large global corporations. As everything becomes more interactive, the more interactions you have, the larger the role of products and services in shaping your impressions toward them.

Understand Users

Your users will likely fall into two categories: the "focused search" people, who want to find specific information or perform a task quickly; and the "loose search" people, who are often killing time by browsing around for familiar or new things that they can later take action upon.

You should enable both types of users to achieve the tasks they set out to do. For the focused search users, streamline processes so that they can achieve tasks in the least amount of steps (or screens) and minimize the additional "gimmicky" features that don't help them. For the loose search users, have broad information that they can browse through. In addition, high glanceability, and easy-to-understand information are key.

Your users expect to have the right information and set of actions when they need them. Mobile apps are very often event driven, so immediacy is key; they run into a friend or acquaintance on the street or conference and quickly need to find a great place to take a group of people for dinner and book a table right away, or to transfer money, or exchange information on which speaker is speaking in which room when. When you are designing your app, think "right here, right now."

Sections of this chapter were contributed by Jez Harper, Tús Nua Designs; Yasmina Haryono, Fjord; and Jennifer Hiley.

CHAPTER 8

Personal Privacy Is Paramount

Consumer purchasing trends have evolved over the decades, moving from a shop owner recommending items to you, to you ordering goods from a catalog on an analog phone, to tracking your grocery store purchasing habits through a register, to cookies tracking your behavior over the Internet, to a unique identifier in your phone tracking everything that you say and do. The amount of consumer metadata (data about data) that is transmitted on a daily basis is astronomical, opening endless opportunities for developers to provide a unique experience to each consumer, and for marketers to be extremely targeted in reaching consumers based on the detailed content and data about the consumer available to them.

Protecting the Consumer

The amount of information you can learn about your consumers when they use your app is astounding. You can gather information on payments, passwords, financial data, e-mail, text messages, purchase history, contact information, pictures, and so on. It is scary to think about how much of this consumer data can be leaked.

How Is Information Stolen?

The most common method for developers to gather consumer data within a mobile app is through Apple's UDID or Google Play's Device ID. The UDID or Device ID is a unique ID that is associated with each consumer device and provides the ability to track the consumer across each app that is downloaded or app interaction. Apple and Google go to extreme lengths to ensure that the data surrounding these unique identifiers is extremely secure, but as hackers become more and more sophisticated, there are opportunities for this data to get hacked. Devices are becoming increasingly vulnerable to the threat of malicious code and malevolent applications. Smartphones, because they are essentially mobile computers based on software platforms, are emerging as a special concern.

Attacks by the Numbers

Juniper Networks (*www.juniper.net*), for example, released a study that revealed a record 400 percent increase in Android malware. *Malware*, which is short for malicious (or malevolent) software, is software used or created to disrupt how a mobile device operaters or gain access to personal information stored on the device. Malware includes computer viruses, worms, Trojan horses, spyware, adware, and other malicious programs. And it's not just Android users who need to worry. The study warned that all device users who download apps are at heightened risk of attack.

ALERT

Another security risk is text messaging. Some 17 percent of all reported infections came from SMS Trojans (viruses), which sent texts to premium rate numbers, incurring major charges to the unsuspecting users.

Privacy Is a Pivotal Issue

Clearly, peoples' lives and devices have become inextricably intertwined. Mobile empowers people to capture and consume content; it impacts how, when, and where they connect with friends and family; it increasingly assists them in daily decision making; and ultimately links the physical and digital worlds. However, the future of mobile will not be decided by technology. It will be determined by people's very human requirements for simple, transparent services they can trust.

The Privacy Influence

Yes, people rely on their mobile devices, and specifically mobile apps, to do a lot. However, before someone downloads and uses an app to its full potential, they must be confident that the integrity of their personal assets (bank accounts, personal data, digital content) is protected.

Recent research highlights the pivotal importance of privacy. In fact, privacy concerns are driving most app users (57 percent) to remove apps from their mobile phones, or to avoid installing the app altogether. This is a key finding in a report released by the Pew Internet and American Life Center, based on a telephone survey of more than 2,200 Americans.

FACT

In 2011, TRUSTe (*www.truste.com*), an online privacy solutions provider that helps organization comply with evolving and complex privacy requirements, published a survey in which one in three consumers named privacy as their number one concern when using mobile applications. The vast majority of people believe that privacy is an important issue and want more transparency and choice over their personal information.

The results come at a time of increasing interest in mobile privacy. But even mounting privacy concerns can't stunt the growth of the mobile app marketplace. The report reveals that apps overall are becoming more popular. Forty-three percent of wireless users now say they download apps to their devices, up from 31 percent last year.

Deleting apps is just one way consumers are protecting their personal data and assets. The report shows that half of smartphone users have erased their device's search or browsing history, while 30 percent have turned off location tracking.

Valuable Resources

Developers seeking information about privacy policies and how these issues affect their applications can gain a lot of valuable information from publicly available resources. The Cellular Telecommunications Industry Association (CTIA), for example, has published best practices and guidelines for location-based service providers on how to protect consumer privacy associated with those types of services. In the meantime, nonprofit organizations such as PrivacyChoice (*www.privacychoice.org*) and the Future of Privacy Forum, as well as the for-profit TRUSTe, offer a variety of privacy solutions, including privacy policy generators or other tools that you can use to create privacy policies that reflect your specific business needs.

In addition to these types of resources, you can employ some general practices to protect the privacy of your customers and your app businesses as well. Overall, the best approach is to develop privacy policies based on a deep understanding of the end-users' requirements.

Many people have attempted to establish principles of best practice, but so far, an all encompassing, practical approach to the issue has not been reached. The regulatory frameworks developed by government bodies such as the Federal Trade Commission and EU commission do not address the global nature of the mobile apps market. What are required are practical tools and solutions from an independent and impartial representative that can represent stakeholders across the value chain, with consumer interests firmly in mind.

The Mobile Entertainment Forum (MEF), the global trade body for the mobile media, commerce, and entertainment industry, is championing such principles through its work, including the recently launched privacy in mobile applications initiative. The initiative will ultimately deliver a single, consolidated tool designed to support application developers through the practical processes needed to establish transparency and informed consent with their consumers. The web-based interface queries developers on the type of information users will be asked for, such as how it will be used and with whom it will be shared.

It will then provide developers, and the app stores that retail the application:

- A straightforward privacy policy unique to each mobile application that explains to the consumer how, why, and to what end their data is being collected
- Screen mock-ups that display how the consumer will be asked for information by the app
- A privacy-rating icon that can be displayed alongside other marketing information within app marketplaces, giving consumers an at-a-glance indication of how their information will be used.

As technology advances, people will conduct more and more of their lives using a mobile device, and those devices will have an ever-wider range of sensors capable of delivering ever more uniquely tailored experiences. The opportunities to observe consumers' activities and behaviors will grow very quickly, but so too will opportunities for abuse of the individual's personal data and privacy.

ESSENTIAL

In many cases, users don't have enough information to decide whether apps are too intrusive. The practical problem is that the data collection practices are just too vague, and this is where the trouble starts. Tell your users up front and straight out what data about them is being collected.

Gaining Trust Through App Best Practices

For app developers in today's mobile space, consumer trust is one of the foremost business-critical issues, the importance of which cannot be understated. In the age of Facebook and Twitter, where people habitually share potentially sensitive information on a daily basis, consumers have been forced to become increasingly aware of the value of their personal data and have come to expect and demand certain safeguards and behaviors regarding its use.

The success or failure of an app, aside from its creativity, usefulness, or innovation, may now hinge upon whether consumers feel they can place their trust in the service provider. Simply put, consumers will not spend money on mobile apps unless they trust them to handle shared personal information responsibly. Establishing such an intimate relationship between developer and consumer can be a difficult task.

Be Proactive

Ask only the important questions and embed privacy measures throughout the lifecycle of your product or service. Create a privacy policy that explains what data you collect, how you use it, and with whom you share it. Building a privacy policy is an important process, even if you do not believe that you are collecting or using data that would trigger privacy concerns. The more information that you collect and use, the more detailed your privacy policy should be. And don't just cut and paste a privacy policy from another app or website. Start by understanding your app in your own terms, and then do your best to communicate the same to your users.

ESSENTIAL

Failure to disclose material information or a misstatement regarding data use practices disclosed in your privacy policy (or elsewhere), could serve as grounds for government investigations, enforcement actions, and private lawsuits.

Communicate Openly and Effectively

Have a comprehensive and transparent privacy policy covering all of your data collection, sharing, and use practices. Use clear and simple language.

Don't access or collect user data unless your app requires it. If you gather or transmit data that your app does not need for a legitimate purpose, you put both yourself and your users at risk. Advertising may well be a legitimate purpose, so long as the collection and transfer of targeting data is transparent, and users are given options about usage of their information for that purpose.

However, platform and app stores may have their own rules about the collection and use of user information for certain purposes, including advertising. Violating a platform's terms of service could get you in trouble with the platform or app store, or even regulators. Delete data that does not need to be retained for a clear business purpose.

Make Your Privacy Policy Easily Accessible

Don't make users search for your privacy policy; make it prominent and easy to find. Provide a hyperlink to your privacy policy prior to download. This means including a link to your privacy policy from your app store listing, or include a link in the sign-up that appears before users have full access to the app. If the app store framework limits your ability to do this, make sure to include your privacy policy in the app itself.

Place your privacy policy in a prominent location within the app under a "Privacy Policy" menu heading or under the Settings menu. If it's not possible to do this, provide a hyperlink to your privacy policy in a similar location. The link should take users directly to the policy with a minimal amount of click-through. Upon retrieval, the policy should adjust to fit the size of the mobile screen.

Consider providing a short form notice—a notice with a limited number of characters that highlights the key data practices disclosed in the full privacy policy—in your app. Seek to provide users with the information needed in the context, at the most relevant time. Provide a hyperlink to the full policy in your short-form notice. PrivacyChoice and TRUSTe provide excellent (and sometimes free) tools to help you create your own short-form notices for users.

Use Enhanced Notice

Don't surprise users; have respect for context. Use enhanced notice in situations where users might not expect certain data to be collected. A privacy policy is an important resource to help users, advocates, and regulators understand your practices, but it is not the only place you should provide information about data collection and use, especially when you are using sensitive data or using data in an unexpected way. Make clear, conspicuous, and timely disclosures when engaging in this use.

If you condition the use of your app on the collection and use of personal information, educate your users about the trade-off. It's fine to condition distribution of your app on certain data usage, such as sharing personal information with ad networks. However, if your application is a "take it or leave it" deal, make the trade-off clear to users so they understand the exchange. Users may be happy to share their personal information in exchange for your app. However, you need to be transparent and up front in your explanation.

Empower Users

Allow users to choose and control the way their data is collected and used. It's important to obtain user permission before accessing location data. Precise geo-location information is increasingly considered sensitive information. You should only collect and transmit such information when you have your users' clear, opt-in permission.

Provide choices to users at the moment and manner in which the notice would be most relevant to the user, within the OS design framework. This usually means before data is accessed, collected, and transmitted or used. For example, your app's privacy policy should be accessible before users download the app and/or before prompting users to register, create an account, or use social network information to log in to your app.

Secure Your Users' Data!

Always use appropriate and up-to-date security measures to protect user data. All applications that access, use, or transfer individuals' data should be tested rigorously for security purposes and comply with current security best practices. Implementing data-retention policies and security measures will help ensure user data is properly safeguarded.

Server security is just as important as the security of the app itself. Servers are the focal point for hackers, as there's more value in breaking into a repository of personal information than trying to compromise an individual user. There are strict rules for storing credit card info and similar care should be taken for personal data.

It's also important to encrypt data in transit when authenticating users or transferring personal information.

Whenever feasible, you should ensure you are encrypting your users' data, especially authentication information like user names, e-mail addresses, and passwords. Storing unencrypted data puts both you and your users at risk in the event of a data breach. If your application accesses, collects, or stores sensitive data or is a fruitful target for phishing attacks, consider using two-factor authentication, such as confirmation text messages or one-time, application-specific passwords.

Ensure Accountability

Make sure someone is in charge! If you are a one-man shop, then this is your job. This means that you must:

- Review your privacy policy before each app release to ensure that it remains accurate and complete.
- Keep an archive of your privacy policy, and ensure that change notices are appropriately posted for users.
- Confirm your company's rules for who can access data internally to ensure that personal information is only available to team members with a need to see it.
- Answer all privacy-related e-mails and communication
- Stay on top of new developments by following the FTC and other industry organizations.

Remember: Good communication is good for privacy and for your business. Provide your users with the opportunity to contact you with questions, concerns, or complaints. This can be accomplished through a simple form accessible from within your app, a feedback forum, or by providing an e-mail address where your users can contact you.

The Privacy Imbalance

Transparency and empowerment are the keys to app success. Users should know what information is collected, how it is being used, and by whom.

They should then be able to impose limits at any time, or change the way the data is used. Allow users to make their own choices in a way that will ensure a trusting relationship between them and developers.

However, this must not be at the expense of accessible and innovative apps that users want to download and enjoy. Ensuring the issue of privacy and user data does not end up stifling innovation and the development of apps that push the mobile experience forward in new ways means there must be a balance between consumer experience and ensuring full disclosure and consent.

FACT

TRUSTe's 2012 *United States Online and Mobile Privacy Perceptions Report* shows 94 percent of consumers consider privacy an important issue, 60 percent feel more concerned about their online privacy today than a year ago, and 58 percent expressly indicate they "do not like" online behavioral advertising (OBA).

In the current app market, this balance isn't always quite struck, with apps behaving in ways unanticipated or unnoticed by the user. The situation can become even more challenging when information is to be shared with third parties. User data is of course a valuable asset when shared with the value chain, and often a welcome benefit of using certain apps with consumers happy to receive relevant targeted advertising based on purchases they've made in the past. But when their information is sold to a company with whom they have no history, and they start receiving ads for products they don't need or want, this can be immensely destructive.

Currently it's up to individual companies to decide what information to share with the user. Knowing what is necessary and essential to divulge, and when to do so, can be very hard to get right when working in isolation, especially if your competitors are cutting corners. It stands to reason that the industry would achieve more through proactive discussion and coordination. The topic needs to be top of mind for the mobile content and commerce industry in order to develop and implement practical solutions and tools built on broadly established principles around transparency and user control.

FACT

The Information Technology Industry Council is striving to establish principles for interoperable cyber security policies and is hoping to gain support from worldwide IT industry associations. The aim is to make sure governments approach cyber security in a way that provides security and also protects innovation.

Attitudes regarding acceptable use of consumer data and what consumers are prepared to share with apps vary widely according to many factors. Consumers are often more than happy to share certain data when it means a free app or spares them from annoying irrelevant questions every time an app is used. The key element in ensuring the continuing goodwill of users and the success of any app is the transparency and informed consent it gives its users, and these can be very difficult to achieve without ruining the user experience.

Sections of this chapter were contributed by Rimma Perelmuter, MEF.

CHAPTER 9

Testing, Testing

By now, you are well into the app creation process. You've designed the app, developed the app, and are ready to send it out into the market. Don't forget to test it! You'll want to test everything about your app: from its usefulness in the mobile world, to the kinds of problems it solves for mobile users and how the design stands up under different contexts. And, you'll need to continue testing it as you polish and add new features.

The Need for Testing

Failing to adequately test your app has consequences that are difficult to recover from. At best, you'll annoy people and the app will get a poor rating on application stores. (Would you choose an app that has three out of five stars or one out of five stars? Most people will choose the three-star app. A bad release with a string of one-stars is difficult to overcome.) At worst, users will complain about your app on social media and you become the butt of an Internet meme that goes viral, giving you exactly the wrong kind of publicity. Arm yourself against a poor user experience and negative feedback by thoroughly and frequently testing your app and fixing as many problems as you possibly can.

Strategy: How and What to Test

Emulators, such as those provided with software development tools are fine for basic testing, but you need to test your app on real hardware to get the best results. Devices have sensor and touch interaction, as well as network connectivity and other physical aspects that emulators can't reproduce.

Just start using the software on real devices the way your end users would, in the contexts, locations, and situations where they would find themselves. Note any odd behavior, usability issues, slow performance, or crashes, carefully outlining the exact steps to reproduce the problem.

If you aren't sure what to report, keep in mind what electrical engineer and author Bob Pease used to say: "If you see something funny, record the amount of funny." Never talk yourself out of logging an issue. If you find it funny (as in odd), annoying, or weird, a lot of other people will as well.

Watch for Deletable Offenses

Deletable offenses occur when a user installs an app and gets sufficiently frustrated with it that they delete it and move on (possibly to a competitor's app). People usually delete an app because it doesn't do what it claims to, it is too difficult to use, or has poor performance.

ALERT

Deletable offenses are bad enough, but don't let them become rant-able offenses! The last thing you want is for people to hate your application so much they complain on social media. Make sure you note anything that is problematic.

Moreover, people are quite emotionally attached to their devices and apps, and if they have a bad experience with your app that generates negative emotions, these customers may lash out at your app, and your company, in public. It's an outcome you want to avoid.

Test Everything Everywhere

Test on as many different platforms as you can, because there is a lot of variation out there. Test on:

- Smartphones and tablets
- A range of operating systems and versions (e.g., iOS 6, Android 2.3, Windows Phone 8, etc.)
- Devices with diverse screen sizes
- Various Wi-Fi sources
- Diverse cellular networks (e.g., EDGE, 2G, 3G, 4G) and carriers

Each of these areas will yield different results when testing. For example, you can test with a smartphone and a tablet at the same time. It's amazing how an app can behave or appear differently on one or the other.

Remember that a testing platform is a combination of a hardware model, operating system version, and—if your app uses network connections—Wi-Fi, cellular networks, and the carrier.

Test on the Move

People use their devices at home, on the move, or away from home. Think of all the activities associated with each of those areas. At home, the light varies day to night and from room to room, and the network strength can wax and wane, especially if you have your own Wi-Fi hotspot.

On the move, the light changes significantly (indoor, outdoor, bright sunshine, cloudy day) and you pass through various network connections (Wi-Fi to cellular networks and back again). You might use your app while walking, jogging, riding your bike, sitting in a car, or on public transport. All of these contexts provide special challenges for your app.

Away from home can include at work, in the park, at a restaurant, in the shopping mall, visiting friends; the list goes on and on.

Test your app according to where and how people will use it, in different locations and while moving between them. You will find errors when you test in one location compared to another. Or you may find your app very difficult to use when moving, or difficult to see in bright sunlight.

Be Ready to Change Perspectives

To get the most out of testing, especially when you are pressed for time, change your testing perspective. The following is a mnemonic to help mobile testers remember quickly how to generate effective test ideas: I SLICED UP FUN!

I: Inputs into the device

Try different gestures, typing, tilting, or moving the device to activate sensors. Use voice control, syncing with devices, and connect accessories like keyboards or Bluetooth headsets. Make sure it handles the input and doesn't freeze up.

S: Store

To get your app in the hands of your customers, you must submit it to an online store. Review the store submission requirements for each store you want to list it on. Some stores are quite strict, so test your app against the guidelines early rather than later in the process.

L: Location

Some of the most powerful features on mobile devices use location services. Various sensors inside the devices and different technologies such as

Global Positioning support this. Make sure location services are accurate when you stay in one location, when you are moving around, and when you move between locations. Always test the transition between Wi-Fi to cellular networks, and in and out of dead spots.

I: Interactions/Interruptions

Check how your app handles interruptions, such as receiving messages from other services, low battery, or syncing. Find out how your app responds when other apps are running.

C: Communication

On smartphones, see how your app responds to interruptions from phone calls, voicemail, SMS/text messages, and push notifications. On tablets, try video chat, and messaging apps.

E: Ergonomics

Since many mobile devices are smaller, people use them on the move, and they don't always use them in ideal conditions. Apps need to be easy to use. Watch for anything that makes you sore: too much typing, or long and confusing workflows.

D: Data

Enter in different character sets from different languages, and make sure the app doesn't crash or freeze. Not everyone trying your app will speak the language you wrote it in. If your app requires data or media files, try different types and sizes of files and make sure it can gracefully provide error messages if files are too big.

U: Usability

Watch for anything frustrating, difficult to use, hard to see, or that takes too much time to do.

P: Platforms

Test on as many devices as possible. Short on devices? Use the "apps with appetizers" model. Organize a gathering with friends who have different devices, and get them to test basic scenarios with you over food and drinks. You'll be surprised at what you find.

F: Functional Properties

Use product documentation to make sure the advertised features and selling points actually work as advertised.

U: User Scenarios

Imagine three different users of your app. Now come up with a compelling story for each of them and a scenario of something they would do. Start with the "technophobe," perhaps a friend or family member who always gets viruses on their computer or can't remember their passcode on their mobile device. What would that person struggle with? What irritating questions would he or she have? Test your app in the way they would use it. You'll find usability problems right away.

N: Network

Wi-Fi and cellular networks are not as fast and reliable as wired high-speed Internet. And, mobile devices can change between network types. Make sure your app handles different network speeds and transitions between them.

What to Watch For

By now you've heard the old saying, "If at first you don't succeed, try, try again." This goes for testing your apps as well. Some problems may come out of nowhere, but many issues are relatively standard. The following is a list of problems to watch out for:

1. Usability issues: anything that threatens the ease of use of the app
2. Crashes: the app disappears, or gives you a crash message
3. Poor performance: the app is slow to respond
4. Strange messaging: error messages that you don't understand
5. Data corruption: the data you save isn't there or becomes mixed up/garbled
6. Upgrades overwrite user settings: updating the app on the store wipes out everything the user saved

Combine Activities

For the best results, pick several ideas and combine them. Type while moving the device around to see how it handles simultaneous typing and sensor interaction. Submit a form while transitioning from Wi-Fi to cellular. Make sure you use the application while doing something else: waiting for public transit, watching television, or sitting in a restaurant. Try it in different temperatures, and outside in different lighting conditions. Also try out the app when your device is in special modes, such as dealing with low battery, poor wireless reception, or if the device is hot.

ESSENTIAL

Get creative to maximize your chances of finding important problems. The more problems you find before releasing the app, the happier your customers will be with your app and your company.

App developers and companies have a lot of time pressure on projects, and the customer expectations for a quality app are high. To make the most of your time and efforts, test early, test often, and test in the real world.

Crowdsourcing for Constructive Feedback

There are billions of users with mobile phones across the world. Some of them are professional software testers, and of these, some work for professional outsourced testing service companies such as uTest (*www.utest.com*) and Mob4hire (*www.mob4hire.com*). They can test your application quickly and relatively inexpensively, compared to maintaining a larger dedicated software testing team. These services can augment your other testing, but not entirely replace formal testing, so factor that into your equation.

To get good results you will need to devote some of your time and effort to defining the tests you want them to run, and to working with the company to review the results.

Why Use Real People?

There are two different aspects of your app you want to test. One is to make sure the app works. In this test you want to find out if the app is functional and does what it's supposed to do.

ESSENTIAL

Functional testing is all about making sure the app works. Will it crash? Is the right information coming up? Will the app download on a specific handset? These are just some of the questions the crowd can help you answer. How can you get started?

Create almost scientific conditions for the test, lay out what you want the testers to do, and the order they should do it in. For example, open the app, log in, click on a specific icon, and so forth. Remember: These testers are savvy app users, so tap into their ability. Let them play with your app and see if they can break it. This will also provide you greater insights into performance and usability that you can factor back into your app development.

The other test is to be sure people, your customers, will like it. After all, just because your app doesn't crash, doesn't mean people will automatically like it. If people do not like it, they will delete it from their apps, or just not download it at all. Mob4hire did a recent study of their testers and found that over 70 percent of those testers surveyed would not even consider downloading an app unless it was rated four stars or higher on the app store.

Think about this. Think of another marketing channel, like apps, that people get to rate right at the source. Can you give a TV show four stars out of five on the TV? How about a magazine advertisement or article? Can you give a radio ad five stars out of five? That is what your users get to do with a mobile app. So how do you get a four-star or higher app? It is not easy, but the solution is clear. Ask real people before you launch your app. Ask them to use the app and ask them how much they would rate the app. If you get 75 percent of the people looking at your prelaunched app and giving you two stars, listen to them. Fix the app, and go through the process again. Keep in mind, emulators or professional testing houses will not tell you how many stars you will get.

Testing for Local Flavor and Audiences

Location is important if you are marketing to the world. One of the most overlooked parts of the testing infrastructure is testing your app under real-world conditions on the actual mobile operator networks around the world. Take the example of an app that uses text messaging to send notifications to the app user. An app may work, but if it takes too long for the text message to be delivered due to local issues in the operator network or other technical issues, then the app may not be usable. How long are your users willing to wait for a text message notification from your app? Well, ask the people in the crowd.

Testing to Tackle Fragmentation

There are thousands of makes and models of devices in the hands of billions of users around the globe. It's not the sheer number of devices that is a concern for app developers. It's the fact that there are vast differences between the makes and models in terms of screen sizes, speed, memory, accelerometers, cameras, Wi-Fi capability, keyboard layout, language, operating system—the list goes on and on.

If you had to buy all those devices, you would go broke! But the crowd has those handsets. Use the crowd.

ESSENTIAL

AppStori, a platform for mobile enthusiasts and entrepreneurs, takes a different approach by connecting app developers with consumers, allowing an exchange around feedback and beta testing that can help developers build a following around their app prior to its release. Cofounder Arie Abecassis established the company to increase transparency around how apps come to market. As he put it in an interview with Mashable.com, "We wanted to allow consumers to develop relationships with the developers, everything from product feedback and data testing."

Mobile Case Study

Case in point: A major retail company came to Mob4hire with an app they were not convinced was as good as it could be. Mob4hire assembled thirty testers with their own handsets in their own homes to do a twenty-minute test plan and then had them spend twenty minutes playing with the app. After the test was over Mob4hire asked the testers to fill out a thirty-question survey.

The average star rating was 3.7. The retailer took a look at the results and discovered a number of things from the testers including the observation that the app drained the handset battery. The retailer fed this back to the app developers to find the problem and seek a solution. The app developers determined that the app was tapping into GPS all the time, not just when it was required by the app to help the user locate a nearby store. This is what was draining the battery, but it's a shortcoming the app developers knew from a real-life test of the app. The developers made corrections to the app.

A month later Mob4hire tested the fix and asked the same crowd to test the new app. The results were an increase to over 4.1 stars. The retailer launched the new app and the average star rating is now approaching 5 stars. They listened to their customers and everyone was a winner.

Think First

Crowd testing is not free. Testers have to pay for their own handsets and data and carrier agreements. A typical crowd test with five testers would cost less than the cost of one smartphone handset.

The more complex the test the more expensive a test will be. If you want to test in ten countries on every carrier on a huge amount of handsets, the cost of a crowd test may be more than you expect. But just compare that to how much it would cost to gather those handsets and to fly to those countries to conduct the tests in real life.

In the mobile industry you do not have time to wait weeks for results. The sooner you need results the more expensive it will be. Remember you are dealing with real people.

Sections of this chapter contributed by Jonathan Kohl, Kohl Concepts, Inc., and Paul Poutanen, Mob4Hire.

CHAPTER 10

Shop Till You Drop

The growth in smartphone adoption and use, coupled with the expansive ecosystem of shopping and social media apps, is empowering a new breed of "super shoppers." A number of important mobile apps, many of which are location-aware and tuned in to social media, are transforming the way people shop. These apps also show the way to new opportunities at the intersection of location, commerce, and user intent. If you are selling products and services in the real world, then you need to harness mobile apps to enhance the experience and encourage purchases.

The Evolution of Mobile Shopping

Mobile shopping is the new mobile megatrend. Beyond the absolute dollars generated by mobile commerce, having a constant presence on a consumer's device presents additional benefits. ABI Research found that 45 percent of smartphone owners with a retailer-branded app visited the store more often and 40 percent bought more than those without the app. Apps act as commerce vehicles, engagement mechanisms, and critical components of a marketing strategy.

Then and Now

It started out as mobile commerce and officially made its debut in 1997, when Coca Cola vending machines were installed in the Helsinki area in Finland that accepted payment via SMS text messages

Mobile-commerce-related services spread rapidly in early 2000. Norway launched mobile parking payments. Austria offered train ticketing via mobile devices. Japan offered mobile purchases of airline tickets. Mobile commerce, and increasingly mobile shopping, is now part of daily life, driven by the advance of mobile apps and consumer adoption. Now consumers have the ability to purchase everything from a cup of coffee to a car, from a sandwich to a time-share on a jet, with their mobile phone.

FACT

Mobile commerce in the United States is predicted to quadruple by 2017, according to the *Forrester Research Mobile Commerce Forecast, 2012 to 2017*. This equates to 40 percent of U.S. mobile phone owners buying products or services on their device and generating $45 billion in sales by 2017.

Understand Your Mobile Shopper

With the increased attention that apps are receiving from all players in the commerce ecosystem, it's important to understand the consumers that are behind the usage.

Shopping Statistics

Research firm Wave Collapse (*www.wavecollapse.com*) conducted a study of 1,000 smartphone owners in April 2012 to identify trends among consumers who are using apps while they are in a retail location to research, and even buy, products.

As with anything around mobile, providing context is important. To do this, Wave Collapse analyzed where people were purchasing products in the previous week. Across the various smartphone and tablet channels (apps and websites), approximately 25 percent of people reported making a purchase on a device. Compare this with 87 percent having made a purchase in a physical store and 60 percent making a purchase online.

What is interesting is that with all the hype about mobile taking away from in-store purchases, Wave Collapse isn't seeing purchases from physical stores decreasing. As a matter of fact, mobile apps could actually be helping trigger purchases. (A very telling statistic about the impact on physical stores is coming up in the section on in-store app users.)

Knowing that 25 percent of smartphone owners made a purchase via their device is a great baseline to have, but another element to build the profile of the mobile shopper is how often they are making these purchases. In this case, Wave Collapse found that of those making a purchase on either mobile or tablet, between 50 and 60 percent had made multiple purchases via these channels. This is on a par with physical stores (69 percent multiple purchases) and higher than online (46 percent multiple purchases).

What the Numbers Mean

What this means is that smartphone and tablet purchasing behavior is different than online. A smartphone or tablet isn't just a device with a different size screen; the experience on these devices drives shopping. Specifically, it's the immediacy and richness of it that makes shopping on mobile devices more akin to the experience of shopping in a physical store. Keep this in mind when you craft your mobile commerce and shopping strategies.

Consumer Motivation

Whether a consumer enters a store, launches an app, or opens a webpage, he or she is driven by one of two motivations: to complete a task (purchase) or to gather information (browse). For the majority of store types, people go in with a purchase mentality. However, for department stores, malls, and electronics stores, people report that they are mostly browsing when they go into these physical locations. This directly impacts the type of strategies that retailers should be considering when offering their app experiences.

Stores Influence Behavior

If you have a store where consumers are mostly browsing, your app should emphasize discovery, allowing shoppers to find exciting items they will genuinely appreciate as they check out what's new.

Discount stores, grocery stores, warehouse stores, convenience stores, general stores (Wal-Mart, Target), hardware/home improvement, automotive, and drugstores are generally about more task-based experiences. People are going into these stores to make a purchase, not to browse. In this case, retailers should be offering apps with shopping assistance functions like shopping lists, coupon organization, reviews, and more. Help them get their tasks done first and then put a bit of discovery in the mix to build engagement.

Comparing physical stores to digital stores (online, smartphone, and tablet), shows that physical stores are mostly purchase-based visits rather than browsing-based visits. However, the opposite is true for online and mobile shopping. People visiting these digital retail sites are reporting that they are more likely to be just browsing rather than going there to make a specific purchase.

ESSENTIAL

Retailers need to embrace their customers' browsing behavior in the digital realm and build experiences around discovery. At the same time, they should not ignore the commerce side of the equation and need to provide an easy purchase experience across all platforms, not just their app.

Preferred Venues

Wave Collapse found that online shopping is the most enjoyable, followed by shopping in a physical store. Tablet shopping, whether in app or in a browser, is slightly less enjoyable. Shopping on mobile phones, again via app or browser, is the least enjoyable of all the channels. That's not to say that mobile is a bad experience or even a poorer experience. It may just be that the larger screen size makes shopping on a tablet a better experience.

It's a different story when it comes to what shoppers are doing in-store. This is where mobile apps are a real crowd pleaser. The Wave Collapse research reveals that about a third of smartphone owners say they use apps to help them shop while they are in a physical store. These in-store app users are a very attractive audience for brands and retailers on multiple fronts. When it comes to purchasing, this group was significantly more likely to have made a purchase, across all store types.

They are going into stores with different intent than what is seen in the general smartphone population. In-store app users are more likely to go into physical stores to browse (remember, the majority of people originally answered in the Wave Collapse survey that they were going into these stores to purchase); and are more likely to use online, phone, and tablet to purchase instead of just browse.

Numbers to Think About

Additional differences continue to paint an interesting picture of the in-store app user demographic:

- In-store app users are more likely to find shopping on smartphone, tablet, and computer enjoyable.
- They are more than twice as likely to have their device in their hand while they are in a store.
- They are also more likely to have made a purchase of a physical product or reservation on their mobile device.

All these findings taken together lead to one conclusion: In-store app users are just shoppers at heart. They like to shop, they shop often and across a variety of locations and devices. The mind-set of this group is going to be savvier and more sophisticated, and that's why they are using apps in-store.

Preferred Features

What features and functions does this app-savvy audience appreciate most? Again, Wave Collapse has some insights you might want to factor into your own app. The majority of in-store app users are using some sort of bar-scanning application. However, they are also drawn by social/reward apps (ShopKick, ShopSavvy, Foursquare) as well as grocery store apps (Kroger, Aldi, etc.). Connect the dots, and there is room for more features and an untapped opportunity for branded apps from grocery stores, for example.

In-store app users are using these apps for three main tasks:

- Price comparison/savings: checking prices as they shop to find the best offers and bargains
- Make shopping easier: taking the hassle out of finding what they want on the go
- Payment: removing the friction from buying what they want

Overwhelmingly, the price comparison/savings features are why people are using mobile apps in the first place.

Wave Collapse asked people who aren't using apps in store what an app would have to have in order to be useful for them in-store. The order of importance matched the task list above. With one exception: "Make shopping easier" was significantly more important to this group than those who use apps in-store currently.

The people that aren't using apps in-store currently are less savvy shoppers, find it less enjoyable, and therefore would welcome features and functions in an app that make the experience easier and more enjoyable for them. Wave Collapse found that even though people may not have made a purchase on their mobile phone, they are still using it to look at products (59 percent report doing this). People may actually be using mobile to "showroom," that is, narrow down their choices before they make a purchase via some other format. Therefore it is critical for retailers to have a good experience on smartphone and tablet in order to be a part of that purchase decision process.

Purchase Patterns

In terms of what is being purchased on mobile devices, hotels, clothing, airline tickets, food, movies, and books are all on the higher end of the products being purchased. Interestingly, food is much more likely to be purchased via smartphone rather than tablet. This speaks to the immediacy and "out of home" usage of mobile phones for this case. The fact that travel categories (hotel and airline tickets) are so high is also telling for the immediacy and "out-of-home" usage of these purchases.

Make sure your app strikes a balance between enabling planned purchases and encouraging the impulse buy. When Wave Collapse asked users if the last purchase was planned or impulsive, many purchases for different categories of goods fell into the impulse side. Although things like travel (hotel and airline tickets) were more likely to be planned, things like clothes, food, beauty, and pet items were higher impulse buys. Take this as another confirmation that your app will likely be well received if it enables users to browse and discover products on a whim.

From this research, it's clear that mobile apps are an important part of consumers' lives, both when they are in-store and when they are shopping in general. Providing an integrated, complementary experience across channels—and within your mobile app—will allow you to enjoy richer interactions with your customers.

Location Is Key

Because consumers have the expectation that they can access anything, anytime on their mobile device from wherever they are, it is important to evaluate location as a key element of your mobile app strategy.

SoLoMo

The confluence of social, location, and mobile—SoLoMo, as venture capitalist John Doerr of KPCB refers to it—is making data available to consumers and businesses that can improve the process and outcomes for both. It represents the growing trend of targeting consumers based on their current location and is typically designed to be shared via social networks. For example, smartphones and a new breed of powerful mobile apps, many of which are location aware and tuned in to social media, are empowering consumers and transforming the way in which they shop.

As consumers engage in SoLoMo (tweeting about their experience, "checking in," searching for nearby businesses, etc.), they are also generating digital signals that are important to businesses. For instance, a large percentage of Facebook posts and tweets contain opinions, both good and bad, about products, brands, retailers, and service providers. For their part, companies are connecting with consumers via mobile channels (apps and the Internet), integrating location-based services into their strategies, and "tuning in" to social media.

ALERT

By mining the digital signals produced by SoLoMo, app developers (and other companies in the ecosystem) can develop a 360° view that allows them to better understand consumers, monitor their performance, and adjust their strategies accordingly.

While there has been a proliferation of apps that combine social and local, including apps from Foursquare, Gowalla, Google Latitude, and Facebook check-ins, there has been controversy over the collection of user data, coupled with concerns over the intrusion of companies into people's personal lives. Some of the new services have disappeared, while the established players (Google and Facebook) have had limited success in converting check-ins into consumer change. Foursquare is a notable exception, which has grown steadily in markets where the online population is drawn to social networks.

Location-Based Services

Location-based service (LBS) can be defined as a social, entertainment, or information service, enabling a company to reach and engage with its audience through tools and platforms that capture the geographic location of the audience. The delivery mechanisms used for LBS include mobile Internet, mobile applications, short message service (SMS) text messaging, multimedia messaging service (MMS), services using GPS, indoor location services, out-of-home, digital signage, print media, and television. According to a 2011 Juniper Research report, location-based services will reach $12.7 billion in 2014.

Location Requirements for Your App

Perhaps it's obvious, but determining the user's location is a prerequisite for any location-based service or app. However, there are a number of alternative methods, each available from numerous providers, that can be used for location determination, and each method and provider has its strengths and weaknesses. Different location-based applications also have different requirements for location determination. The following examples explain the role that location determination plays in location-based apps, as well as ways in which location determination requirements vary by app.

What You'll Need

Do you want to localize your content for delivery to your customer? You can use services such as MaxMind (*www.maxmind.com*) to look up location based on a visitor's IP address, which will allow you to tailor your content and include "local" news, weather, and offers. Because IP lookup provides only a crude measure of a user's location (identifying the end-user's location within twenty-five miles about 85 percent of the time), you may want to look into other services that narrow this down. Skyhook (*www.skyhookwireless.com*) has introduced Loki to make it easier for users to reveal their exact location to websites.

Are you focused on delivering an app that allows users to stay connected to friends and followers nearby? Foursquare, Loopt, Where's Beacon Buddy, and myriad similar services provide a useful blueprint. They allow friends, family members, and associates to share their location with one another, making it easier for them to "hook up."

Location is also at the core of a variety of other scenarios. Here is a sample of the use cases and services that evolved to meet the need of users in these situations.

- **Find objects:** Described as a "compass on steroids," GeoVector's World Surfer allows you to "bookmark" the location of your parked car and can even help you find your way back to it.

- **Navigation:** With GPS and real-time LD, NDrive, Google Maps Navigation, and a host of other providers can give drivers precise, turn-by-turn directions.
- **Alerts:** Services such as LoJack can detect when a vehicle or laptop has been moved beyond a "geofence" and issue an alert to the owner and authorities.
- **Health:** GE, Intel, and others are developing solutions that use mobile technology to monitor the movements of elderly individuals in their homes. Inactivity can prompt a call or visit from a caregiver.
- **Shopping:** By tracking shoppers' movements in stores, micro-location-based solutions can determine whether a particular point-of-purchase display has caught shoppers' attention.
- **Security:** Using RFID bracelets, Safe Place allows hospital staff to monitor newborns and determine whether an infant is in the nursery or with the mother. In each example, the location of a person or object must be determined. However, the requirements for location determination vary across apps in terms of precision, frequency of refresh/updates, and other aspects. Clearly, one size location determination doesn't fit all location-based apps.

Leveraging Location

"Location, location, location" has always been the motto, but today, it is especially important for retailers that are striving for true loyalty from their customers via mobile. Location-based marketing allows brands to engage with consumers anytime, anywhere, providing retailers with opportunities to reach them at—or just moments before—the point of purchase.

The Power of Personal

Because of mobile's personal power and immediacy, it provides unique and effective ways to deploy timely and time-sensitive messages and incentives. In addition, proximity and presence technologies allow brands to significantly enhance customers' experiences by personalizing the information and incentives they provide based on their location. Consumers want information that is relevant to their lives—when they need it.

With today's connected consumers, loyalty (and in-store sales) is incumbent upon reaching consumers *at the perfect moment*—not five minutes after they have left the store or just after they made a purchase (which is when most retailers today start tracking customers). It is when they are in stores and *thinking* about making a purchase that matters.

FACT

With an estimated 55 percent of smartphone owners having used their devices for location-based information such as restaurant recommendations or driving directions in 2011, it is no secret that these technologies could transform the consumer experience.

While there are numerous consumer benefits that result when retailers leverage proximity and presence marketing, here are a few that, when used properly, can go a long way in establishing app loyalty and trust.

Tips for Establishing Loyalty

1. **Incentives.** Give customers incentives such as coupons or special offers to keep them engaged. When you know exactly where your customers are, you can send coupons or special offers right before they make purchases or are en route to shopping—both scenarios affect consumers' purchasing decisions and help establish loyalty with apps.

2. **Product knowledge.** Shoppers are shopping with their mobile phones. If brands know customers are seeking certain items (based on previous conversations with them, or what they have learned from past purchases), they can send them more information about relevant products, enticing them to buy. For example, sending a product review allows consumers to quickly access information via their mobile phones in order to make them feel confident in their purchase.

3. **Reducing time and distance.** If you know customers' location, you can provide them with helpful information and deals when they are literally right near stores, reducing the time and distance they need to go to find their desired purchases. This helps to combat "showrooming," when

consumer's comparison shop from their mobile devices while they are in-store.

Examples of Markets Leveraging Location

Of course, mobile shopping is hot, but it's not the only market where the power of location can transform experiences. Location-based services aren't simply limited to deals and check-in services; they can be applied to a broad range of businesses, from travel to retail, restaurants, and education.

People are social animals and have a natural desire to connect with other people—and form networks and communities—based on factors including common interests, shared passions, and proximity.

FACT

The app makers behind Banjo have harnessed SoLoMo. Banjo's connection engine taps into the most popular social networks (Facebook, Twitter, Foursquare, Instagram, and more) to provide a real-time view of what's happening anywhere in the world. Since its initial launch in June 2011, Banjo has grown to more than 2 million Banjo members in over 190 countries.

Location-based features are also popping up in dating services everywhere. Users can take advantage of the GPS capabilities of today's smartphones to show nearby singles that meet their dating criteria. Apps can make recommendations, enable people to digitally flirt, explore users' profiles on other social services like Facebook or LinkedIn, and of course arrange to meet up. Several such location-based apps already exist, such as Skout, and Blendr, which allow users to network with other locals and check in at venues using their phone's built-in GPS.

The music industry is being reinvented via location-based storytelling. The app Broadcastr records and shares stories in audio format and pegs each one to a specific location. Users can search for stories by location or category or may opt to "follow" a person whom they consider to be a good storyteller, sorting stories by that person into a special tab. Listeners can rate stories as they hear them, and stories can be shared with others via e-mail, Facebook, or Twitter.

A similar platform is offered by German start-up SoundCloud. Sound-Cloud is a social sound platform where anyone can create sounds and share them everywhere. As users create these sounds, they can tag them to the places they were created via built-in Foursquare integration.

Coming at it from a slightly different perspective is Pandora Radio. Because Pandora collects zip codes, marketers can use geo-targeting to reach specific demographics through the system. The system enables advertisers to target users by age, music genre, and geography.

Entertainment

Social platforms can also extend and enhance TV shows, much in the same way that readers' comments do for news articles. Brands now have the ability to use the knowledge of a viewer's location, time of day, and other demographic data to create a mobile response to what is seen on TV, triggering a call to action that drives traffic to websites or physical retail locations.

One example of this was from the 2012 Super Bowl, where one-third of all the ads were connected to the mobile app Shazam. When the Shazam-enabled TV commercials aired during the broadcast, the app responded by displaying a mobile ad and a location-based offer with directions to the closest retailer for redemption.

Politics

Utilizing LBM not only allows potential voters to easily get up-to-date on what politicians are doing, it also makes it easier to identify like-minded voters with whom they can connect and share ideas.

Geopollster is a good example, a service that works with Foursquare so that users can "cast a vote" during election periods for whichever party they support every time they check in somewhere. Buildings are then displayed on a map as leaning toward one party or another.

Monetizing Location

From location-aware shipping helpers, to location-based marketing schemes, apps that are at the intersection of social, location, and mobile are gaining serious traction. But how can app developers make money?

Tracking Lost Phones

Just about everyone has lost their phone at some point. The tracking apps available to smartphone users provide them with reassurance that if they ever happen to accidentally misplace their phone, they can quickly locate it. Once users have installed a tracking app on their phone, the location of the phone is pinpointed through the use of GPS, Wi-Fi, and cellular triangulation techniques.

These services represent a revenue opportunity for carriers and other third-party service providers like Loc-Aid to license technology to service providers.

Finding Work/Finding Stuff

If leveraged properly, location-based services can offer the extra boost that helps people find the people and things they need, close by. Several platforms have emerged, such as Zaarly (*www.zaarly.com*) and Task Rabbit (*www.taskrabbit.com*), that are focused on posting services and things needed by location. Sort of like a mobile and location-based Craigslist, these companies take a commission on the rental of a product, completion of a task, or sale of an item.

Pay-per-Call

Since the mobile device is, after all, a phone, the progression from high-intent-driven search, to qualification, to placing a phone call is a natural one. Pay-per-call is something that will therefore find a home in mobile much more than it did on the desktop.

Google and others have already gone down this path, while showing strong performance for both call and click rates when localized content (including phone numbers) is present in ad copy.

Deals/Discovery

Another monetization model involves what Foursquare provides for free; a dashboard for businesses to launch and manage local deals and loyalty programs. The goal here is to drive trackable promotions and dynamic demand generation.

This opens the opportunity to verticals where there's time sensitivity or perishable inventory (great if you have a restaurant). It then straddles local advertising and yield management, especially when bundled with things like scheduling tools, as Groupon has developed.

FACT

A recent study from the Location Based Marketing Association (LBMA) found that 39 percent of consumers said coupons for nearby stores was one of the most appealing aspects of location-based advertising.

Mobile Gaming and Location

Location services can help address the two biggest challenges most game developers face: increasing user engagement and maximizing revenue. One company already succeeding in this is Rovio, the maker of Angry Birds. Most players of the game complete the free levels and never buy the paid version. To combat this, Rovio created Magic Places by partnering with retailers like Barnes and Noble. The company even announced that players could take their Nook Color into a Barnes and Noble store and use the Mighty Eagle for free to clear levels in Angry Birds. The Angry Birds Nook app costs $2.99, but there is no cost for using the Mighty Eagle in stores.

This kind of partnership can become a great method for retailers to drive additional foot traffic and sales, while at the same time helping gaming companies that rely on in-app purchases to grow their revenues through retailer subsidy.

Sections of this chapter were contributed by Asif Khan, Location Based Marketing Association (LBMA); Joy Liuzzo, Wave Collapse; Phil Hendrix, immr; Gary Schwartz.

CHAPTER 11

App Stores and More

App stores are a great way to reach potential customers and users of your apps: They offer relatively easy access to a huge range of people, and they handle the billing and delivery of your products. But while they may provide a nice storefront, they don't always do a lot to help market your apps, and competition in the biggest stores gets more and more difficult as the number of apps on offer via their storefronts continues to increase.

Choosing Your App Store(s)

The avalanche of mobile apps has caused an irreversible change in the content distribution landscape. The first wave of mobile content—ringtones, images, and wallpaper—was delivered via specially focused portals run by mobile operators. Handset makers like Apple were nowhere in sight, and Android didn't exist. Fast forward to 2012 and the mobile app stores are the primary go-to-market channel for well over half of mobile app developers. In fact, Vision Mobile's *Developer Economics 2012* report reveals that mobile operator portals are the choice for just 3 percent of the 1,500+ app makers surveyed.

FACT

Leveraging the absence of the Google Play app store in China, China Mobile's app store has a customer base that amounts to 22 percent of its subscribers, and has served over 600 million downloads, according to *IHS Screen Digest*.

The Apple App Store, the proprietary software store launched by Apple in July 2008, resulted in the sales of billions of apps for the iPhone, iTouch, and iPad devices. With numbers like these, it is clear that app stores are the place to be. But before you decide where to offer your app, explore your options.

The Wireless Industry Partnership (WIP, *www.wipconnector.com*), a global organization connecting developers to information, resources, and people, counted over 125 app stores. From the Amazon app store, which allows you to market your apps to tens of millions of Amazon customers using Amazon's proven marketing features and manage your apps using convenient self-service account management tools, to GetJar, the world's largest, cross-platform app store, which counts 50 million downloads a month across more than 200 markets, to MiKandi, the "world's first adult-only mobile app store," you're sure to find a fit with your target demographic.

Of course, you need to be focused. Submitting an application to an app store is oftentimes more difficult and time-consuming than you'd expect, and this burden, added to the ongoing management of each submission, means it's likely that you will want to choose a small number of stores to work with, rather than trying to hit them all. So how should you choose? WIP divides your options into the following categories.

Major Platform and Niche Market Stores

Major platform stores remain the most likely place users will go looking for apps, so even if your outside marketing may focus on other stores, it's worth putting your apps in these. Does your app have a strong draw to a particular market niche, based on geography, type of content, or even device type? For instance, if you have an app for Android tablets, there are stores that focus on them; if your app appeals to bookworms, you might want to check out the Barnes and Noble Nook store; if you are targeting urban youth, you can target three primary stores:

- **GetJar:** Publish a Java-based app and distribute alongside 350,000 other mobile apps
- **Nokia Ovi Store:** Publish a Java-based app and distribute to Nokia's feature phone models
- **Blaast:** Publish a Java-based app and instantly distribute it over the cloud for 2,000+ devices (from Nokia to BlackBerry and local brands). From Blaast's cloud-based platform apps and updates are instantly distributed to all Blaast users. It's free to publish apps on Blaast's platform. Blaast's apps consume up to 80 percent less data than web browsers and native apps, which makes them affordable for most users and ideal for busy network conditions.

These smaller stores often attract more dedicated users—and offer up a smaller pool of applications and have stronger marketing outreach.

Operator Stores

Many mobile operators now support their own app stores. While the *Developer Economics 2012* report shows this is a hugely popular distribution channel, there are significant benefits depending on your business model. Keep in mind that mobile operators take the hassle out of charging for your app by offering carrier billing, which has been shown to have a dramatic increase on app sales. Many mobile operators also offer marketing support in the form of featured placement, advertising on operator websites, and other collateral benefits. And exposure on a mobile operator app store goes a long way if you want to reach users outside your home market. Many operators from smaller markets are now actively recruiting applications for their stores and are willing to help with converting them to local languages and marketing them to their user base.

FACT

Google Play is delivering 1.5 billion apps a month as of 2012. At the Google I/O conference, the company announced that its Google Play store now offers 600,000 applications and delivers more than 1.5 billion installs per month, totaling 20 billion app installs so far. Developers will also be able to make use of Google Cloud Messaging, which allows the free and unlimited delivery of notifications to Android users.

The Major App Stores

Clearly, the mobile app store landscape is in a state of flux. However, here are some of the key players to know and watch:

- **Apple:** For most developers (and many in the tech press), Apple's App Store is the number one attraction. The ability of Apple to control such a significant slice of both market share and mindshare is baffling if you consider that its ecosystem is even more closed than its operating system. Put simply, Apple controls all software distribution through its App Store. However, the choice of payment mechanism (credit card) effectively bypasses mobile operators to make monetization for developers a no-brainer.
- **Google Play:** While it does offer significant volume, developers continue to struggle when it comes to monetization. The problem: payment—or rather a lack of it. The lack of suitable payment mechanisms has incentivized other providers—including Amazon and GetJar—to launch alternative storefronts offering Android applications, and so tilt the ecosystem in their favor.

- **BlackBerry App World:** BlackBerry offers downloadable apps for BlackBerry device holders through its app store titled "BlackBerry App World." BlackBerry is putting significant emphasis on its BlackBerry 10 operating system and opening it up to developers. This is one new operating system to keep an eye on in the coming months.
- **Windows Phone Store:** With Windows Phone's upcoming launch of its Windows Phone 8, their new app store will offer developers updated tools and functionality in its new SDK. This is an important development for Microsoft as it invests significantly in its new OS.

FACT

A new feature of Microsoft's Office 2013 allows associated applications to be purchased from within the program itself, and an API will allow developers to create HTML5 and JavaScript apps for Office products.

Submitting to the Major App Stores

The major app stores are operated by handset makers or companies that, like Google, are stakeholders in an operating system. And there are many arguments for only submitting your app to these major players. One of these app stores might be the perfect match for your target audience.

Why It's a Good Idea

If you are targeting a device such as Apple, then you know you can best get your app in front of these users if you submit your app to the app store where they all congregate in the first place. It's not that these users have much of a choice: Their device and the app store that offers apps for that device are inextricably intertwined.

Another reason is the numbers. If your app is on multiple stores, and it is getting downloaded or purchased by users across these multiple stores, then the information around your app—including ratings, reviews, and number of downloads—may not give customers a clear picture of your app's popularity.

For example, you may have had hundreds of thousands of downloads across all the app stores, but if any particular store is only showing 500 downloads, your app will appear as relatively unpopular and this may sway the user to choose a competitor's app.

The end user is not the only concern when it comes to statistics across multiple app stores. You, as the owner or developer of the app, also need access to accurate numbers so you have a clear picture of how well your app is really doing in the marketplace. This is where a one-stop, holistic view of your app success, in terms of downloads, ratings, reviews, and other data is critical.

Of course, there are online tools, such as AppAnnie (*www.appannie.com*), you can use to bring order to the chaos, but these tools will only retrieve data from the major app stores. If you need the data from alternative app stores, you will need to collate it yourself, which will be time-consuming, and prone to errors in calculations.

There are also valid arguments for pursuing a strategy to place your apps across a variety of app stores. Maybe you want to reach niche audiences, such as enterprise app users, or markets abroad where the mobile operator, not the handset maker, is the main distribution channel.

ESSENTIAL

Don't be afraid to look beyond the Apple App Store. After all, a sharp focus on Apple ignores the massive uptake of other platforms and the popularity of other app stores. However, you need to consider the payment options available in each of the stores so that you are not limiting your potential audience.

Whatever your reasons, just be aware that app store fragmentation can be a headache. From marketing assets to billing engines, there are vast differences between app stores.

This is where companies like CodeNgo (*www.codengo.com*) are rising up to fill the gap. Specifically, the start-up CodeNgo is focused on streamlining the complicated, and often painful, submission process. In 2012, the company launched its first service, allowing developers to submit to multiple app stores and manage their developer accounts all from a single location. The aim is to save developers time and ensure that their apps are available regardless of where consumers choose to shop.

Tackling the App Approval Process

Different app stores have different review and approval processes. It's outside the scope of this book to cover them all, but here is what you need to know to offer your app via the Big Two: the Apple App Store and the Google Play marketplace. A major difference between the two giants is the review processes, or the length of time it takes to review your app and release it to the world.

With Google Play, it takes no time at all because your app isn't reviewed by anyone. The Apple App Store can take anywhere between days and months (!). Granted, this has only happened on rare occasions, but you should be prepared. Your business depends on this, so factor in a waiting period. Overestimate the period of time it will take and you'll never be disappointed.

History and experience shows it will typically take Apple ten days to review your app, although most of this time your app is simply in a long line waiting to be reviewed. The actual review time is between one hour and a few days. Keep in mind that these timescales are not just for the first time you submit your app for approval. The amount of time you wait will be similar, maybe even slightly shorter, when you submit updates to your app.

Tips to Speed Up the Process

Apple has managed to remain secretive about many of their business processes, and developers commonly complain that their app review process takes place in the black box. Because so little is known about how the review process works, it follows that it's difficult to offer any tips to speed up the process.

However, there is one way you can at least potentially speed up your app review: Ask Apple to expedite it. There is a difference of opinion as to whether asking for an expedited review results in a faster review time, but experience shows that it is more likely to happen if it is for an update to fix a catastrophic bug, and almost certainly won't happen for the first submission of an app.

Many developers have gone on record saying that they have been successful asking for an expedited review, but a significant number argue it didn't help at all. Since there is no consistency in the review times, it's tough to say if the request for an expedited review really resulted in a faster review time. It may be that these developers just got lucky.

So, there are no surefire ways you can speed up the review process. But there are a few steps you can take to at least ensure that you are not making the review process any longer than it needs to be:

- Read and follow the app review guidelines: Many people are amazed when their app is rejected because of a breach of one of the sections of the app review guidelines. Ensure that you are aware of everything in the app review guidelines and check that your app is compliant.
- Test, test again, then test some more: There are quite a few apps that get rejected because only the "success" case was tested. In other words, during testing, the app was only tested to ensure that it did what was expected when used as expected. This is confirmation, not testing. The app needs to be tested to ensure that it works, even when it is used in an unexpected manner. The app also needs to be tested to ensure that it fails gracefully and keeps users informed every step of the way. Say your app requires an Internet connection and none is available. If your app crashes without informing the user, then your app will be instantly rejected.
- Leave notes for the app reviewer: If your app requires certain information, such as a login, or if it is complex, then leave a note for the reviewer. The more you can do to make the reviewer's job easier, the better chance your app has of being accepted.

Getting the Go-Ahead, or Not

You've submitted your app for review, and waited for the response, probably impatiently, checking the progress on a regular basis, and finally, you've received the e-mail telling you how your app fared during the review. The e-mail will either tell you that the app has been accepted and can now be published on the app store, or it will tell you that the review raised one or more issues that need to be corrected and will provide details of the issues.

If the app has been accepted, and you selected the option to release the app as soon as the review has been completed, then your app will be on the app store. If you selected the option to choose when to release the app, you can go onto the app portal and release it whenever you want. The latter option may make sense if you want to rev up the marketing machine first and release your app with a storm of planned and timed social media blasts, guest blog posts, and an old-fashioned press release to mark the event. What to do to make the most of your app premiere should be baked into your marketing and promotion strategy from day one.

However, the bigger question is what to do if your app is rejected. There are many reasons for rejection, some that you hadn't considered, some that you thought you could get away with, and some that you disagree with.

So, do you change the app to remove the reason for rejection, or do you argue your case? Many developers have tried to argue their case and convince the app store provider that they made a mistake. But, before you decide to take this on, be aware that there are very few cases where the app developer has come out the winner. Your chances will depend on the level of change that you would have to make to ensure compliance, and what sort of impact that would have on your app. Ideally, you should investigate all other possibilities before deciding to argue your case.

Understand and Harness Momentum

Many of the top download charts are really based on momentum, rather than sheer number of downloads. This is because app charts look at what apps users have been downloading over the past twenty-four hours, week, or another period of time, in an attempt to remain fresh and fully up-to-date.

If you want to ride that wave, then you have to get ahead of that momentum. Some developers tried to upload a new version every six weeks to always be displayed in the "new" sections—but this loophole didn't last long. Instead, work toward a big burst or spike in downloads over a few days, with the aim of achieving extraordinary downloads for three days, then waking up with a much-improved rank on the fourth.

Just like Google has a secret algorithm for search, every single app store out there has its own formula to determine your ranking. Don't waste time trying to crack the code. Accept that the main factor determining the rank across most app stores seems to be the number of downloads in the previous three or so days (or similarly short period of time). How else can you create momentum around your app? You may want to start off with a sale and discount your price for a few days. Like retail sales, you may want to pick an event that extends the appeal of your app. Think back-to-school if your app is educational. Halloween is a great time for "grisly" games or novelty fun. And don't forget the holidays when people are in the market for apps so they can have fun with the phone or tablet Santa brought.

Encourage Customer Reviews

Following these tips is a great way to get your app in front of your audience. But you also need to pay attention to customer comments and feedback if you want to be sure your app continues to be a crowd pleaser. Sure you may see and celebrate your reviews in the app store, but don't rest on your laurels. You will need to take action, and be creative, since no app store will allow you to respond to app reviews directly.

You see a positive review of your app and you want to say thank you? Well, you can't. Even worse, you see a review that trashes your app. The reviewer claims it's useless or just doesn't work, a review that obviously deters potential customers from even trying to download and use your app. You want to point out that the review is completely off the mark, and that your app works just fine, but again: You can't. Or maybe you just want to delete an old review that complains about a bug you have long since fixed. Again, you can't edit the comment to show the issue has been resolved.

So, what can you do to get around the complete lack of a meaningful feedback loop between you and your customer? In a word: nothing. But you can follow some helpful tips below to equip your app users with the next best thing.

- **Provide instructions:** After you've spent a huge amount of time designing and developing your app, using it will become second nature to you. As a result, you might assume that your app is so simple to use that no one will need instructions. Wrong. No matter how simple you think the app is to use, provide instructions, either as in-app instructions, a user guide on your website, or a YouTube video, or any combination of these.

- **Provide feedback mechanisms:** If your app crashes, or if it behaves in an unexpected manner, or if it is so confusing users will want to tell you about their frustrations, either so that you'll correct the issue or just to have a rant, and if their only outlet is to leave a review on the app store, then that is what they will do. This isn't what you, as the app owner or developer, want them to do. You want them to tell you directly so that you can correct the issue, or correct their thinking, so that they will be a happy user. Provide the mechanism for your users to send you an e-mail directly from the app to give you feedback.

- **Swing the balance in your favor:** Face it: You are more likely to get reviews from users who are experiencing problems than from those who love your app. A great way to encourage a more balanced view of your app—and collect more balanced reviews in the process—is to ask the user to rate the app with a simple "thumbs up" or "thumbs down." That way, you have a clear and instant indication of whether the user is happy or not. If they are happy, then ask them to leave a review or a rating on the app store. If they are not happy, then enable them to write and send a message to share precisely what you need to do to improve their experience with the app.

App Store Shortcomings

App stores may have reach, but they lack facilities—or focus—to help developers market their apps. The *Developer Economics 2012* survey reports app stores rate only average in terms of helping developers promote, price/bundle, and target their apps.

More importantly, app stores won't let you connect with your customer after the sale to get feedback, or any other information to help you hone your app, cross-sell, or up-sell features or just improve your promotion. In fact, most app stores obscure customer account information, making it difficult to understand.

Facebook is determined to create value for developers with the Facebook Platform and the Facebook App Center. As of June 2012, developers can now determine which devices a user has and customize their app or game experience accordingly. In addition, Facebook will also begin allowing advertisers to target promotions based on their customers' e-mail addresses and phone numbers. Businesses will be able to use their own collection of customer e-mail addresses and phone numbers to target ads, using a method that Facebook officials said was secure and would help companies reconnect with customers.

Which App Stores Keep Their Developers Happy?

What is the measure of a healthy app store? Size matters, but it's also important that the app store provider cultivates a robust ecosystem, where developers can prosper and profit. Mobile analytics company Bango recently took the temperature of the global developer community with a survey that garnered more than eighty responses. Granted this is not a huge sample, but it does shed important light on how developers rate important areas including app submission and payment processes.

How satisfied are developers with the ease and speed of the approval process? Respondents overwhelmingly preferred the Android-based Google Play approval process, with 73 percent saying they were "very satisfied." When it comes to the approvals process, BlackBerry App World came out on top, with 97 percent of developers reporting they were either very satisfied or satisfied with the overall process. Significantly, Apple's App Store fared worst of all, but still saw 78 percent of developers either very satisfied or satisfied.

How do developers feel about the payment models available, such as subscription and in-app purchase? This question revealed that most developers are content with the range of options available, with more than half of respondents calling themselves "happy" with the methods and models at their disposal. Only the Amazon App Store received lukewarm feedback from developers, with just over half of respondents reporting that the range of payment options for developers was "okay."

How satisfied are developers with the revenue settlement (payment) process and statements? Here BlackBerry App World came out on top, with 37 percent of respondents saying they were "very satisfied" and 55 percent said they were "satisfied." Google Play received the lowest satisfaction score, with 16 percent reporting they were either "dissatisfied" or "very dissatisfied."

FACT

There will always be developers who simply follow the money, and with this in mind the app measurement company Flurry conducted research to find out what developers make on top-ranked apps across Apple, Amazon, and Google Play, across a forty-five-day period. Among the findings: For every $1 a developer earned on an app distributed via the App Store, an Amazon app store user would secure $0.89 and a Google Play user made $0.23. These results are more remarkable when considering that the Amazon store is Android content, just like in Google Play.

Payment

A hard look at the total payouts to developers from January 2011 to November 2011 tracked and reported by the research firm Piper Jaffray shows that Apple paid out a total of $3.4 billion. In comparison, Google paid out $240 million. BlackBerry appears to fit somewhere in the middle between Apple and Google. The company has publicly stated that BlackBerry App World generates thirty downloads per user per year, and that those downloads generate 40 percent more revenue for developers than Google Play. In fact, 13 percent of BlackBerry developers have made more than $100,000 from their BlackBerry Apps.

A Store for Enterprise Apps

While there is certainly excitement surrounding apps aimed at consumers, the real opportunity may be in developing specialized apps designed to meet the specific needs of large and small enterprises seeking to support their workforce with apps that boost productivity and drive positive business results.

What is the size and scope of the enterprise app opportunity? In a word: huge. Big companies are hungry for business apps. In fact, nearly four in five large companies would like to purchase mobile applications for various business uses, according to a survey of company execs conducted and published by Partnerpedia, a mobile app development services company.

Don't expect them to buy apps from popular app stores managed by the likes of Apple. The business community takes issue with purchasing business-tool apps through these consumer-focused stores for a variety of reasons. The top reason is a lack of business focus, a concern named by nearly 58 percent of respondents. Other concerns focus on the inability to own or control app licenses and the potential problems associated with security (or rather a lack of it). Additionally, organizations also cited a lack of volume purchasing or PO purchasing as another reason why they would prefer not to purchase apps from consumer-focused stores.

ESSENTIAL

Already analysts observe that enterprise app stores are wrought with challenges and confronted with issues ranging from app certification to neutrality (because the software vendors who manage them are loath to promote similar apps from industry competitors).

Sensing a business opportunity, major software vendors are hoping to follow in Apple's footsteps and open up their own so-called enterprise app stores, which sell not only their own software but software and services from an ecosystem of partners and resellers. It should be noted that the term *enterprise app store* is used in this case as a vendor-hosted electronic marketplace serving up apps to customers. This shouldn't be confused with another meaning of *enterprise app store*, whereby a company serves up apps (usually mobile ones) to employees.

Made to Order

What are the apps enterprises want most? Here is a list to help you focus your efforts and resources:

- **Management:** apps that help companies conduct paperless meetings or take notes securely.
- **Marketing:** apps that help the company connect with its customers and explain/demonstrate key value services, products, or broad-brush value propositions.
- **Sales:** apps that offer presentation tools equipping the mobile work force to sell to customers on the spot.
- **Finance:** apps that enhance business cash flow or allow staff to input time.

Set Your Priorities

Do you want to get more bang for your buck? Think about a multiple store strategy in order to reach a larger market. Do you want to be 100 percent sure you only target users whose device can download and run your app? Then you most likely want to stick to the major app stores.

Whatever strategy you choose, decide what matters most to you—reach, buzz, support—and build a matrix of app stores that meet your requirements. Stick to these criteria in prioritizing your app submissions. But don't do this in a vacuum.

A good way to understand the dynamics for each app store is to consult (directly, via blogs or forums) with other developers to learn about their experiences. This will help you weigh the benefits as you map out your app store distribution strategy. Other elements you may want to factor into the equation include: the number of app store users, the number of competing apps on offer there, the monetization methods, the revenue share you can expect on app sales, the application process, and the overall user experience the app store offers.

If you feel confident about your app, then go it alone. There are several white label options available now where you can have your own app store on your own site. The tools offer the functionality of a full store for a customer to download and pay for your app.

Sections of this chapter were contributed by Jez Harper, Tús Nua Designs; Andrew Bovingdon, Bango; Chris Jones, CodeNgo, Heini Vesander, Blaast; Mike Anderson, the Chelsea App Factory.

CHAPTER 12

App Monetization Models

The avalanche of mobile apps turns up the pressure on individual mobile developers to find new ways to rise above the noise and generate meaningful revenues. A great app is a good start, but developers determined to build a serious business also need a solid strategy. Finding the right match between business models and monetization methods is core to competitive advantage.

Monetization Models Explained

Now that you've developed your app, you have a lot of app stores and storefronts where you can sell your app, but you also have to choose a payment method and mechanism wisely. A singular focus on credit cards, for example, could exclude important customer segments, such as digital youth, who don't have credit cards but do have a huge appetite for apps.

You should also be aware of the increasing importance of billing solutions that give you greater control over the merchandising experience. These solutions allow you to sell content, virtual goods, add-ons, and updates right from within the app using one-time payments or an ongoing subscription scheme.

App developers have a wide range of options to choose from when it comes to generating revenue, and your choice will depend on your business model, scale, and target market. And keep in mind you will need to be flexible. From one-off payments, where the developer charges up-front for an app, to product placement, where a brand pays the app developer to appear in the app (much like marketers pay Hollywood studios to make sure the new James Bond movie is chock-full of their clothing, cars, and cool gadgets), to per-unit royalties, where handset makers and platform owners pay app developers for exclusive app distribution.

Pay-per-Download

This model, also known as the one-off payment model, is very easy to understand. You charge the end-user for downloading and installing your app via a one-off purchase on the app store. Then you pay the app store provider their cut of the revenues and keep your share.

If you have a good app, then sell it, but it's worth considering the app store refund policy if this is your chosen route. Some app stores don't process the payments until twenty-four hours after the app is downloaded. This allows the customer to try the app and return it a day later if they're not happy. It also allows them to use the app or play the game for a day without paying.

In practice, most of the leading developers have found that paid applications underperform other types of monetization. Why? Typically the consumer aversion to paid apps prevents them from trying the app. A notable exception to this is Angry Birds, which started out by selling its flagship game for ninety-nine cents. If you're going to charge up front for your app, ninety-nine cents is a great price point.

Freemium Model

Freemium means that you offer your app free of charge to the user (to download and use in its most basic form). If a user wants to have an even better experience, they can pay for it. The goal here is to get as many users as possible to the pay threshold as quickly as possible by enticing them to go further with the experience.

Distribute a free app with limited functionality then use in-app billing to charge users to get the pro version. Charge to add more features or levels in a game. Introduce your own virtual goods within communities. This gets users familiar with your app, and when they are ready they can pay to get more. It focuses the payment on the place and time the customer is most likely to upgrade, avoiding loss of sales that are sure to occur when passing customers back to an app store to download the premium version. Freemium also lends itself to viral marketing, rewarding users that spread the word with extras that would normally require a purchase.

How do you get your users to open up their wallets and make that all-important purchase decision? Here are some helpful tips:

- **Tease them.** Provide enough in the free version of the app to build interest. How do you know if you have an app that will keep users coming back, and paying, for more? One way is to measure user retention (whether they return again to the app). If new users download your app and never use it again more than 60 percent of the time, you need to update the core delivery. Keep in mind that you shouldn't ask for money too early in the game. If opening up the app runs them up against a pay threshold (that is, you charge them to pay if they want to continue), they will most likely abandon and never return.
- **Be clear.** The incentive to pay must be easily understandable and completely transparent. Your app must state exactly what the user will get when they pay for the extra content or features and give good reasons why they should do it in the first pace. If you are not sure if you are doing this right, then ask a test group, or just a bunch of your

friends to read your terms, and they will quickly validate whether the proposition is clear.

- **Make it seamless.** If the user has to leave the app to go off and acquire the currency/points/virtual rewards to purchase what you offer, it will fail. You do not want to have a pause in the action when the user is ready to dig into their wallet and make a purchase. Show them the options then and there, and be sure that you return them to exactly the point when they decided to pay.

- **Deliver on the promise.** Users intrinsically evaluate the ROI. If they pay, they should have an edge over those users who didn't. Be sure to offer them more for their money. And keep tuning what your app offers to reach higher levels of repeat purchases per monetizing user.

- **Innovate.** Don't stop adding layers/levels/rewards to your app. Space them out and make them invaluable and your customers will likely continue to pay for them. This is a value exchange. Get it right and people will even pay more.

Subscriptions

The subscription model is quickly becoming one of the more popular models for mobile applications, particularly among apps in the news/publishing genre. Just like in real life where subscriptions are sold to print media like magazines and newspapers, apps in this space charge for fresh content and exclusive information.

If your app is a service delivering new content on a regular basis, subscription billing is a good way to maximize your revenues. If you can keep users engaged and get them to subscribe, you will have recurring revenue each month in addition to your new user revenue.

Here are a couple of tips if you're thinking of going with a subscription model:

- **Keep the app fresh:** Each month or even week, there should be new videos, books, songs—whatever niche you're in. However, if you're doing a service, having a reason for the customer to justify paying you every month is crucial. Statistics have shown that consumers are willing to pay on a recurring basis, but only if they see the value in it. So be sure your customer service and content shine.

- **Pay close attention to app store policies:** Perhaps one of the biggest hurdles in the subscription model is the complicated relationship between you and the app store(s) you choose to partner with. Each app store has its own unique policies when it comes to revenue sharing, and some stores have rather strict guidelines when it comes to subscriptions. Check before you make your users promises you can't keep.

- **Partner with the operators:** You may want to work with mobile operators since they can also promote your app to their customers. After all, the subscription model is what made the first wave of digital content—ringtones, wallpapers, and single-track downloads—so incredibly popular. Having a relationship with an operator—who already controls the customer relationship they have with their subscribers—can lead to a lot of exposure for your app, and guaranteed subscriptions.

ESSENTIAL

Ask yourself: Does your audience have a disposable income it will spend on apps? If the answer is yes, then a paid model might make more sense. If the answer is no, then it may be best to focus on offering your app support by advertising.

In-App Payments

As mobile app sales and distribution across all major platforms continue to flourish, the need for mobile billing solutions that give developers and app store providers complete control over the experience has never been greater. Specifically, these solutions allow app developers to sell content, virtual goods, and add-ons directly from within their app.

Against this backdrop, the advance of in-app billing, which allows developers to collect one-time payments or start ongoing subscriptions within their app using a variety of payment methods, opens up an exciting range of commercial opportunities.

What's more, in-app billing enables developers and application store providers to deliver customers a simple and consistent application payment experience on operator networks as well as over Wi-Fi.

ESSENTIAL

As more and more platforms, players, and app developers enter the market, in-app payment solutions will become more attractive. This is also the view of market research firm Juniper Research. Its *Mobile App Stores: Business Models, Strategies and Market Segmentation 2010–2015* report forecasts that in-app billing will increase in use, enabling incremental revenues.

App Stickiness Matters

According to mobile analytics and marketing research company Localytics (*www.localytics.com*), creating app stickiness is important when trying to generate repeat app visits and subsequent in-app purchases. In other words, a smartphone user who engages with an app for several consecutive days is more likely to make an in-app purchase than an infrequent app user.

FACT

Free mobile applications comprise 89 percent of all downloads, but in-app purchasing is helping developers monetize them, according to a Gartner report. App store revenue from in-app purchases is expected to grow from 10 percent in 2011 to 41 percent by 2016.

A study by Localytics shows that loyal app users generated 25 percent more in-app purchases than average customers. Moreover, Localytics data showed that 44 percent of smartphone users who made an in-app purchase did not do so until they had interacted with the app at least ten times. It also found that users who made in-app purchases did so, on average, twelve days after first launching the app.

In 2011, Localytics found that 26 percent of apps were used only once after download. This further indicates that user experience and engagement are crucial for generating repeat app usage and subsequent in-app purchases. Mobile marketers should observe the mobile gaming market for best practices around the in-app business model.

How to Make Money from Your Apps

With millions of apps available across a number of platforms, gaining your slice of revenue from this crowded marketplace can be a challenge, but never fear, you can reap the rewards if you approach it in the right way.

How you monetize your app will depend on the platform you have decided to write it for. On one hand, you have the flexibility of the latest HTML5 platforms currently being heavily promoted by many businesses. On the other hand you have native apps, known and loved by all the operating system and platform manufacturers because their high performance locks consumers to their platform.

Monetizing on Mobile

The payment experience plays a major part in the number of consumers who successfully pay. Minimizing the number of payment steps and the amount of data that needs to be entered by the consumer significantly increases sales.

This is why operator billing delivers such high results. Even new customers can place a charge on their phone bill with a single click, without the need to register or enter any details. Operator billing can deliver over 70 percent success rates, even for new customers. Traditionally the payout rates achieved from operator billing have been lower than credit cards; the best payout rates available reach around 93 percent, but most operators are lower. Fortunately, this is quickly changing, but there is still a premium for truly frictionless payments.

Credit cards require the entry of the card number, expiration date, and security code, all resulting in significantly lower conversion rates than operator billing. This is especially true for new customers, where conversion rates are typically lower than 15 percent. But once the card details have been provided, they can be reused to deliver high sales success, often higher than operator billing, exceeding 80 percent (credit cards have higher maximum spending limits and are less prone to insufficient funds often found with pre-pay mobile accounts).

Popular online payment solutions like PayPal simply add another level of complexity, requiring up-front account registration and repeat login each time the customer wishes to pay.

When it comes to integrating these payment methods, you have two options. Firstly, if you are a big brand with significant time, money, and resources, you can negotiate directly with the operators and card providers to integrate directly. Alternatively, you can use a third-party mobile payments platform that already has direct connections with a wide range of operators and card providers. The latter is the best route. Not only does it remove the cost and complexity encountered with direct connections, it dramatically improves time to market and can deliver significantly higher sales conversion rates.

The Price Is Right

Before pricing your content, make sure you do your homework and don't make your app too cheap or too expensive. Try some different prices, offer promotions during quiet periods, and reward social promotion with discounts. Also, make sure the price works for your target market in all territories. Unfortunately, one size doesn't fit all. An app priced for the United States will be too expensive in many countries. Don't just rely on currency conversion rates; check the local market values.

Taking the Pulse of Popular Monetization Models

While the price of your app may vary, one thing is key: To capitalize on the mobile applications market, you must ensure consumers can effectively discover and seamlessly pay for your apps. Experiment with different payment models and make sure you present a good payment experience to increase the volume of successful transactions. Also track and measure how users discover and use your applications, so you can continuously improve the performance.

Business Models for Game Apps

When your game or app is ready to release to market, there are many options you can choose to make it big in the app economy. Currently freemium games are the hottest model, but you can't just follow the trend—it's important to carefully consider which business model makes sense, based on the attributes of each game. There are a wide variety of strategies, including ad-supported, premium, or hybrid models. Before diving into choosing the right model for your game, here is an overview of the four models game developers use most frequently.

1. Premium—Premium is defined as charging $.99 or more for an app or game. Premium was the predominant model early in the app store and it is used by some of the biggest names in the market like Angry Birds, Doodle Jump, and Cut the Rope.
2. Freemium—As mentioned before, freemium is the practice of giving your app or game away for free and making money via in-app purchases. Freemium games currently make up thirty-two of the top fifty grossing apps on the iOS App Store including: Dragonvale, Rage of Bahamut, Poker, and Bejeweled Blitz.
3. Ad supported—Ad-supported apps are, well, supported by advertising. Ad-supported games typically include banner advertising or other types of in-app advertising. Some of the more common types of in-app advertising are:

    - Banner advertising: Banner ads enable app developers to display small advertisements throughout the app, or on specific screens.
    - Video: Video ads enable app developers to show short (fifteen- to thirty-second) videos during natural breaks in an app.
    - Offer walls: Offer walls are typically used in freemium games and apps. They enable developers to provide users with free items, like currency, in exchange for taking an action such as signing up for a free Netflix trial, liking an app or product on Facebook, or other related tasks.
    - Interstitials: Interstitial advertising, like video, enables app developers to show full-screen ads at natural break points in a game or app. In most cases, these ads market other third-party applications.

4. Hybrid: Of course, you can implement any of the models described above in various combinations. The two most common hybrids are: freemium with ad-supported and premium with in-app purchases. When developing freemium apps with consumable items, the apps typically display offer walls or incentivized video views. If you are planning to release a title supported with banner advertising, it is a best practice to include an in-app purchase item to remove advertising. Some example apps that use this strategy are Memory Matches 2 and Paper Toss.

ESSENTIAL

Ad funding is a great way to make money. But it can be a risky business. Ad networks promise a lot, but check that they actually deliver. The benefit is that they are open to all to download, but you will only earn money if your app gets run and the ad is clear and compelling.

Choosing the Right Model for Your Game

There is no shortage of options and nuances within each business model. Making your initial choice is incredibly important to your game's potential success. Here are two tips to help you decide which business model is most appropriate for your app:

- **Content first.** Put what's best for your content ahead of what's best for your monetization. If your model doesn't fit your content, you'll confuse consumers, leading to low retention and revenue. Say it out loud: content first, money second.
- **Bake it into your app.** Actually, you need to decide how you are going to monetize your app early in the development process. The absolute worst thing you can do is force a business model onto the app *after* the app is complete. Decide which model you are using at the beginning, and then build your app accordingly. If you're going to use banner advertising, you need to build space for the banner into the app UI. If you are utilizing interstitial or video advertising, you must design natural breaks in game play to display interstitials (video, rich media,

and so on). If you're going to use in-app purchases, the game design must incorporate purchased items and allow users to purchase the items at appropriate times and levels.

Virtual Currency to Consider

If your game includes consumable items, it should be distributed using the freemium model. The most important tool for freemium app developers is in-app purchase. In-app purchase allows app developers to sell packs of content or consumable items at specific price points. When using in-app purchases (IAP), it is important to test often and adjust pricing as needed to maximize revenue and end user engagement. If you are just getting started with in-app purchase, look at the top-grossing apps that utilize this method to get a basic overview of how in-app purchase should be implemented.

As freemium has become a more popular model, developers and publishers are adapting variations of the model for different types of games with fantastic results. Casual puzzle games can use the freemium model, if implemented and optimized correctly. For example, Bejeweled Blitz, one of the biggest brands in casual gaming, has maintained a top-fifty grossing rank for months.

Selling Games with the Premium Model

The rise of freemium does not mean that the premium model is not a viable business model for games and apps. In fact, many popular game apps have successfully utilized the premium model to entrench themselves in the top 100, building very profitable businesses along the way.

How do you know if your app is best served as a premium game instead of freemium? Start by asking yourself if there is a natural way to integrate consumable items into your app, like rewards, additional features or content, or virtual currency. If there really isn't, then you should consider a premium model.

If you are using premium, then it's helpful to involve your customers in the process. To accomplish this you should recruit app advocates. This means finding a way to enable effective viral marketing. If your consumers like your app, they will want to share it with their friends.

ESSENTIAL

Make it really easy to post scores to Facebook and Twitter. Make it really easy to invite friends via e-mail and social networking sites. Reward users for sharing your app with some form of free content. No matter which model you choose, you need your consumers to do your marketing for you. What about adding an ad-supported strategy into the mix? The decision between premium and ad-supported is easy: Do both.

It's well documented that it is substantially easier to get users to download free content vs. paid content. Providing a free version of your app or game is always a good idea. The next question becomes: How do you convert those free users into paying users? There are two methods to convert free users to paid users:

- Provide a limited, free version of your app that includes a strong upsell to the full version. Use analytics to understand user behavior and maximize conversions to the paid version of your app.
- Provide a full, free version that includes in-app advertising and provide an in-app purchase item to remove those ads.

Of course, there is a segment of the consumer base that will never pay for an app, so providing them with a full experience that includes in-app advertising enables you to earn some revenue. Those users that do not like advertising can pay to remove it. The secondary benefit with this option is you can focus your marketing efforts on one version (free) rather than having to market two versions (free and paid).

Sections of this chapter were contributed by Andy Bovingdon, Bango; Ryan Morel, PlacePlay; and Dan Appelquist, BlueVia.

CHAPTER 13

Mobile Marketing and Advertising

Mobile is the only industry in the world where you can create a product, offer it for free, and potentially make more money than if you charged for it in the first place. If you are looking to grow your user base substantially, advertising is one of the only effective approaches. By some estimates, an app distributed for free will have ten times more downloads compared with a paid application.

Mobile Marketing Defined

With the growth in smartphones, mobile websites, and apps, mobile has become the fastest growing category in advertising. According to eMarketer, mobile advertising spending in the United States will grow 80 percent to over $2.6 billion in 2012. And many experts expect mobile advertising to exceed desktop web advertising in the next few years. Whether you are an app developer, publisher, or marketer, mobile advertising is a hot topic.

The mobile marketing landscape has evolved significantly over the last ten years. From mobile messaging through to push notifications on smartphones, app developers have many new and burgeoning opportunities to target consumers today. Mobile marketing and advertising used to be very tightly controlled by the mobile operators, which insisted on strict guidelines marketers were forced to follow to run these campaigns, sometimes taking weeks at a time just to program and run a campaign to target people on their mobile phones. With the advent and rapid adoption of smartphones, consumers now have a multitude of channels to receive communications from brands and marketers. One of these is your app.

The Power of Mobile Marketing

As the mobile marketing opportunities open up and evolve, it is important to first identify your marketing goals and then select a channel for your mobile marketing program. There are a number of ways to target consumers as of this writing and there will be ten more by the time this book is published. Here is a narrow list of current opportunities to choose from:

Mobile Messaging

SMS (short messaging service) allows a mobile user to send and receive a text message of up to 160 characters and across virtually any operator network. This service is also referred to as "text messaging" or "texting." All mobile phones shipped over the past few years support SMS. As a result, the large installed base of SMS phones creates a large addressable market for SMS-based mobile marketing campaigns. You can harness SMS to send text messages to encourage your database of app users to try out new features or explore other related apps. You can reactivate users that you have on record as having downloaded your app, but not on record as having used it in the last weeks/months.

Banner Advertising

Graphical, interactive display ads are the predominant ad unit. In most cases, banner ad impressions can be purchased on a cost-per-thousand (CPM) or a cost-per-click (CPC) basis. Mobile offers targeting possibilities beyond that of traditional media. As this develops further, expect to see a range of targeting options made available covering context, demographic, and behavioral attributes. Any targeting options made available will comply with existing national-level legal and regulatory frameworks governing privacy and personal data. As an app developer and publisher, you can allow their inventory to be sold by a third party, either as premium inventory or as part of a mobile ad network.

ESSENTIAL

Banner advertising is the easiest for app developers because you aren't limited to choosing one; you can test many and find the one that works best for you.

In-App Advertising

Buying advertising in mobile apps is similar to buying advertising on PC applications. For nonrefresh ad serving, where the ad is delivered at time of download, ads can be purchased on a cost-per-download basis. In current and future scenarios, ad impressions can be purchased by CPM, as long as robust and trusted reporting capabilities are in place. Dynamic ads in mobile applications can also be purchased on a cost-per-click (CPC), cost-per-acquisition (CPA), or cost-per-unique-download (CPD) basis; the latter is ad sales based on the number of unique users reached by the advertisement.

There are a number of mobile ad networks to choose from for your display or banner advertising campaign. These include:

4th Screen Advertising

Aarki

Adfonic

AdIQuity

AdMarvel

Admob

Admoba

Adsmobi

Adultmoda

Airpush

Applifier

Burstly

Buzz City

ChartBoost

Hands

InMobi

Inneractive

Jumptap

Kiip

Leadbolt

MdotM

Medialets

Microsoft Mobile
 Advertising

Millennial Media

mKmob

Mobclix

Mobfox

Mobgold

Mobile Theory

MobileFuse

Mojiva

Mopub

Nexage

OnMOBi

Pontiflex

Reporo

SellARing

SessionM

SponsorPay

Startapp

StrikeAd

TapGage

TapIt!

TapJoy

TapTap Networks

Vserv

W3i

Wapstart

Webmoblink

Widespace

xAd

Ybrant

YOC Group

Yoose

Zumobi

Push Notifications

Push notifications allow you to send messages directly to the people who have installed your app, even when the app is closed on a device. Create engagement by delivering relevant information including sports scores, breaking news, stock movements, or game challenges. Send messages to your full audience, segment your audience into specific groups, or send custom messages to individuals.

Mobile Coupons

A "mobile coupon" is an electronic ticket solicited and or delivered to a mobile phone that can be exchanged for financial discount or rebate when purchasing product or service. Customarily, coupons are issued by manufacturers of consumer-packaged goods or retailers, to be used in retail stores as part of a sales promotion. They are often distributed via SMS or MMS, or triggered by interaction with a mobile app, such as using an app or other mobile means to find a nearby store. The customer redeems the coupon at the store or online. In some cases the customer redeems the mobile coupon at the store and the retailer forwards the redemption to a clearinghouse for final processing. What is unique about mobile coupons is that the memory of information in the coupons often outlasts the expiration date of the coupons themselves, triggering actual purchases at later dates.

Location-Aware Marketing

Any marketing that pulls data from the device or the mobile app is considered to be location-based marketing. Always provide clear opt-in instructions and explain as quickly and clearly as possible to your prospects and customers how to opt in to your program. Explain to consumers what they can expect after they've opted into your program. That way, you'll avoid unpleasant surprises for your target audience. Test the campaign on unbiased candidates. Make it worth their while. You're asking customers to opt in to your location-based marketing program, so reward them for their trouble.

Mobile Video and Interstitial Advertising

This is a mobile ad unit that displays a video to the end user from within an application. As you integrate video advertising, you'll need to decide if your app will allow users to close a video before they have finished watching it. If you don't allow this, you'll increase your revenue, but you may experience a decrease in retention as users become annoyed with being required to watch fifteen- to thirty-second videos.

FACT

Keep in mind that as the number of video ad impressions grows (a measure of how many video ads are shown) by orders of magnitude, costs per view (CPV) are coming down. On average, app developers can expect between $.01 and $.05 per *completed* view, so you need thousands of views to make a meaningful amount of money.

Offer Walls

Offer walls enable you to provide your users with currency or other in-game consumable items for engaging in activities such as signing up for Netflix, completing a survey, and so on. On Android, this also includes the ability to obtain items by downloading another app, though this isn't possible on iOS. Offer walls are the ultimate value exchange, but they don't always provide a fresh stream of offers for users to choose from.

There are a variety of offer wall providers on the market, including Tapjoy and W3i. One challenge that developers face with offer walls is that most providers insist on exclusive agreements, making it difficult to determine what works best within your app.

Before you make your choice, talk to other developers who use the offer wall providers you are reviewing. Sure, you might be competitors, but you're all in this together and can benefit from sharing knowledge. Take advantage of the existing knowledge and resources available to you in the developer community.

A Word about Location-Based Marketing

Location-based marketing (LBM) bridges the gap between all forms of marketing media, inclusive of social media, the Internet, out-of-home, and real-life interaction. LBM covers the utilization and/or integration of all media to engage and market to people in specific places with specific offerings. LBM uses location-based services to reach and engage with consumers based upon where they are located.

Understanding the Differences

If location-based services are the technology and media delivery platforms used for the identification of an individual's location and preferences, then location-based marketing is the use of these platforms by brands, retailers, and their agencies to target the message to individuals and engage with them based on their location. The key is that location is most often linked to a specific intent to buy or research products and services at that moment in time.

Brand/Sponsors

The source of funding for the advertising and marketing industry primarily comes from brands and sponsors. Their budgets are traditionally broken down between TV, print, radio, and digital. If the digital budget represents an average of 20–25 percent of the advertising budget, mobile is currently only a fraction of that total.

Mobile app developers seeking to leverage location-based ad dollars need to consider how the data collected from those apps can be used to make traditional media more effective and measurable. The value of location-based marketing for brands and sponsors resides in the information captured about their customers' behaviors and the opportunity to engage with them at the right place and the right time.

In the past, a beverage company like Coca-Cola or Pepsi would rely on restaurants or grocery chains to market and promote their products at the local level. LBM enables them to go direct and interact on a real-time basis. This capability can be used for a wider range of activities including product development and social media promotions based on "likes," consumer reviews, and recommendations.

Mobile Advertising 101: The Basic Lingo

If you are a developer looking to monetize your mobile apps, there are various options at your disposal, but studies show that mobile advertising is a good way to go. Due to the hot nature of the mobile advertising space, everyone wants a piece of the action. There are many mobile advertising companies competing for your business. Relax, this is a good thing. Competition is always good for the consumer.

FACT

Advertising will account for 23 percent of mobile-application revenues this year, an increase from 18 percent last year, according to a Flurry study. Ad revenues from mobile apps will total $2 billion in 2012, more than double the 2011 figure, the study reports, although most revenues will continue to come from paid apps and in-app purchases.

Choosing a Mobile Advertising Company

So how do you choose what company is going to lead your mobile advertising efforts? There are several business models to consider for mobile advertising and over 200 companies to choose from . . . and counting!

Premium and blind mobile ad networks remain one of the most popular ways to promote apps. It's generally thought that, while generating users via mobile ads is more expensive than incentivized networks, the value of the user (the ARPDAU) via mobile ads is higher. There's a multitude of networks out there that provide everything from simple banner ads, right through to video ads and rich interactive ads displayed in mobile sites and apps. Some mobile ad networks are very well set up for promoting apps, and will allow developers to cross-promote apps or games within the network in exchange for running ads. Check out MobyAffiliates (*www.mobyaffiliates.com*) for an up-to-date listing of major mobile ad networks for more information.

Blind Networks work on a CPC (cost-per-click) basis. These networks serve a high volume of advertising largely from independent mobile app developers and offer self-service tools to help advertisers track and optimize campaigns. Blind networks include companies like InMobi, Madvertise, Adfonic, and BuzzCity.

Premium Ad Networks, which include the likes of Mobile Theory, YOC Group, Microsoft Mobile Advertising, and others—tend to focus mostly on brand advertising CPM campaigns and therefore use a smaller number of more premium publishers like big-ticket mobile sites, mobile operators, and top-tier publishers. Premium networks can also offer performance advertising (CPC), but the cost of running these campaigns is typically much more expensive than it would be on a blind ad network. Sometimes CPA (cost-per-action/acquisition) is also available through premium networks.

Premium blind networks fall somewhere in the middle of blind networks and premium networks. They typically have a higher number of premium publishers than blind networks and attract a higher proportion of brand advertisers. Premium blind networks usually offer both CPC advertising as well as CPM. The cost of advertising on these networks varies significantly and is often negotiable. An example of a premium blind network is Millennial Media.

An ad exchange enables cross-platform mobile application developers to maximize revenues with one embedded SDK that serves ads from multiple ad providers. This category includes companies like Inneractive, TapJoy, Mobclix, Smaato, and Nexage. An ad exchange partners with global ad networks and local premium agencies to provide developers a global coverage of local-targeted ads. The exchange provides high fill rates, click-through rates (CTR), and CPM. An exchange also offers developers and brands the ability to directly buy inventory from publishers to run their own campaigns on.

FACT

An app monetization exchange is the next generation of ad exchanges offering multimonetization streams on top of display and rich media ads, all integrated to a single robust SDK, which is offered by companies such as Inneractive.

Getting Started

If you plan to manage mobile advertising on your own, don't put all of your eggs in one basket. Test different advertising channels to see which work best before making too large of an investment in any one network. If you invest the time in getting to know folks at mobile ad networks, you should be able to get a few test campaigns for free or at a reduced rate to trial the network. The costs of advertising can add up quickly, so it's important to time your investments carefully. Advertising in well-timed targeted bursts is usually more effective than advertising a little bit each day over a long period of time.

Integrating Ads Into Your App

Earning a living from an app is no walk in the park. Not every app achieves the success of Angry Birds, Cut the Rope, or Fruit Ninja. Putting aside the issue of getting a mobile app discovered and downloaded, developers, especially on Android, are finding it increasingly difficult to generate any significant revenue from their apps.

App monetization is a huge challenge, and there are nearly one dozen models to choose from, including mobile advertising, as well as in-app purchases. Another route is paid search advertising, which picks up on users' keyword search terms to suggest apps that match.

The following is a list of three quick tips on how to integrate ads into a mobile app in a way that will generate clicks and, ultimately, more revenue for you, the app developer.

- **Relevant and targeted:** If you have advertised on the Internet or even dabbled in the world of traditional advertising, you'll know the name of the game is "conversion." How many people see the ads and—more importantly—how many of them actually become customers. To drive CTR (click-through rates) in mobile advertising, you must be able to deliver the right and most relevant ad to the user. In practice this means making sure a user in Japan is not presented with an ad for a sporting goods offer from a store in New York, for example. Likewise, a person viewing a children's book app should not be served an adult advertisement. Targeting users based on factors such as location and the content they view is essential and sure to increase the bottom line numbers of clicks, conversions, and overall revenue.

- **Helpful and not annoying:** The last thing users want to experience is an ad that prevents them from continuing whatever it is they were doing in the first place. If they are busy playing a mobile game, the ad should not interrupt the game play. Instead, the ad should appear at precisely the moment that it would actually encourage the user to engage with the ad. Displaying an ad between levels, for example, is one of these moments. Thus, the ads you are serving should be relevant, helpful, and implemented properly to avoid annoying the user. Integrate this thinking into your mobile advertising and monetization strategy and you can greatly enhance the overall user experience.

- **Integrated and well designed:** Ads should be perceived as an integral part of the app in which they appear. An ad that sticks out like a sore thumb not only ruins the aesthetic of the app itself; it will most likely *not* generate a click by the user. If the ad appears in the same color scheme as the UI of the app, or if it appears on its own separate screen in between levels, then it is not a distraction. Instead, advertising has become part of the experience, encouraging users to engage and click the ad. This interaction (as opposed to interruption) boosts the developer's revenue—and the enjoyment of the app user. But don't limit yourself to just one ad format. An ad that has an interactive layer that facilitates rich media such as video will be even more of an enticement for the user to click the ad.

ESSENTIAL

Mobile advertising generally requires you to negotiate different commercial agreements with a variety of ad networks and agencies around the globe. This is a grind that requires time, which is equal to money. Don't get defeated, though. The journey is worth it!

Advertising Challenges

The mobile app marketplace has become crowded with thousands of apps available for download. The increase in app consumption has led to a corresponding rise in the number of services available to app developers. In-app advertising solutions make up a large fraction of these services, providing a bewildering array of options for app developers to choose from. To map out the right advertising you need to ask the right questions.

Questions to Ask about Advertising

Ryan Morel over at PlacePlay, a provider of app developer tools to help increase engagement and revenue, suggests your start with the big questions first:

- **Are you providing a premium experience?** Will your app compete on quality with big titles like Angry Birds, Cut the Rope, and so on? If so, banner advertising might not be as appropriate as video, interstitials, or offer walls.
- **How do users progress through your app?** Do users play and complete a level or set of objectives, have a break, and move on to the next level or set of objectives? Or is it a continuous experience? If your app includes natural breaks in game play, it is well suited to interstitial or video advertising during those intermissions.

- **Does your app include consumable items?** Are you using freemium game mechanics? If so, you should be using some form of offer wall in order to provide your users with a simple value exchange: advertising engagement for currency or whatever your app uses.
- **What are the long-term goals for your brand?** This is a big question. For now, keep this in mind: If you plan to be in the app business for the long haul, you *must* think about how you want consumers to perceive and engage with your brand and how you are going to move your consumers from app to app and from experience to experience.

The takeaway: App developers need to take a step back and think about how they are going to approach the market over time. It's a worthwhile exercise that will help you identify the most appropriate advertising solution for your app, your brand (even if it is not yet established), and your business.

Maximizing Your In-App Advertising Revenue

Now that you have decided what type or types of in-app advertising you are going to use, and which mobile ad networks you are going to source inventory from, it's time to maximize that revenue. The following are six tips and techniques that you can apply to increase your in-app advertising revenue. These tips and techniques apply to banners, videos, interstitials, and offer walls.

Ask for User Location Data

Obtaining location data is as easy as a permission request on Android, but requires location-based features on iOS. It's well worth it, as location-targeted advertising can pay up to four times more than advertising that is not location targeted. As mobile ad networks get smarter, location will enable context, interests, and a variety of other factors to be incorporated to help provide the right advertising at the right time.

Don't Overdo It

As with many disciplines, less is more in mobile advertising. How many times have you used an app that showed an interstitial every thirty seconds, or multiple banners on a screen, or any other type of really invasive advertising? There are lots of problems with invasive advertising, but the biggest one is that you are annoying users to the point where they will:

- Start ignoring all your advertising, making it valueless
- Stop using your app, removing any opportunity you have to monetize them, convert them to another app, or get a recommendation to one of their friends

Here are some related rules of thumb:

- Place banner advertising at the top or bottom of your app and don't make it annoyingly easy for a user to accidentally click on an ad. Accidental clicks are frustrating to users and virtually worthless to advertisers.
- Don't show interstitial advertising or videos more than once every three minutes. Anything more frequent than that increases the risk of losing users. It is okay to put up a video or interstitial ad after ninety seconds, for example, but don't show another for at least three minutes after that.
- If you are going to use video, try to integrate it into the experience and make it subtly clear that this is how you are making money. For example, after a user has passed a level you can include a video ad with text along the lines of: "Wow, great job! While the next level is loading, please watch this video from our sponsor."

If you include banner advertising, that doesn't mean you can't use interstitials or vice versa. Some people may disagree with this, but it's really up to you to determine what fits your content. The benefit of employing multiple ad types is that you create multiple revenue streams and have levers to pull in case one isn't working for you. But, remember, don't overdo it.

Reaping the Benefits

Once your mobile advertising systems are set up, the basic formula is quite simple: More downloads equals more traffic, which equals more impressions and clicks, which means more revenue for you. Not onetime revenue, but ongoing cash that your integrated ads are generating for you twenty-four hours a day, seven days a week.

Be Relevant

Due to the nature of the mobile phone and its "always on" capabilities, receiving relevant and targeted ads within a mobile app can actually contribute to the overall user experience. Issues arise when ads appear that are completely irrelevant or are displayed in a way that interferes with the core functionalities of the app. Use precise targeting to ensure your app is relevant to your user. Even better if you can match your app with their context or intent.

ESSENTIAL

When analyzing which advertising networks you want to partner with, make sure to look carefully at how each company keeps its advertisements relevant. This means that selecting the right company for your mobile advertising efforts is a crucial step. You must ensure that your ads can be targeted per location, per app category, and per content.

Relevance can be broken down into three main categories: location, content, and the user. Today's targeting capabilities are endless. A fashion company can advertise its brand to consumers who are using fashion apps. The same goes for games, sports, and entertainment. You can also target by location, which means the user sees advertising linked to their location.

Behavioral targeting adds value to the return on advertising by identifying relevant users for advertisers. The cost of targeted advertising is more expensive, but by reaching the most relevant audience, the ad spend is also more effective. Reaching just the relevant audience is also a better user experience, as users are now being inundated with ads they find irrelevant and dismiss as spam. Mobile advertising is in a similar place that online advertising was five years ago before ad networks, ad exchanges, and ad platforms evolved to incorporate behavioral targeting.

It's not just important to provide relevant advertising your users will appreciate. You should also make sure you provide your users with a simple way to turn off the advertising if they desire. There are two main ways to do this:

1. Provide an in-app purchase option to remove advertising (or make it part of any other in-app purchase). For more details on how to set the right price to remove in-app advertising, check out PlacePlay (*www .placeplay.com*), which also provides some helpful tips you can follow. PlacePlay reckons no more than 10 percent (and that's probably high) of apps that include in-app advertising provide their users with an option to remove advertising altogether for a price. Now, you may say that only a small percent of users will take this option, so why bother? Well—that's okay if you have 1 million users and 3 percent purchase the item, that's a meaningful amount of money—especially if you have the in-app purchase option priced correctly.
2. Provide an upgrade to the full version of your app that does not include advertising. The biggest benefits to doing this are that you create an additional revenue stream and you give your users the choice of accepting advertising, rather than making it solely your decision.

ESSENTIAL

Add a countdown button to the bottom of your ad—but think about delaying the "Skip" button by a few seconds. When presented with the option of clicking a skip button, the majority of users will, due to the clear call to action. Now, you should definitely provide the user with a way to bypass ads if they choose to, but there is no need to have the button show up immediately. Keep in mind that your users could actually benefit from an ad they see through your app, but they intuitively skip it due to the design of the skip button. Delaying skipping an ad by a couple of seconds won't hurt anybody, but it may be beneficial to both users and developers.

Measure, and Measure Again

Operating without analytics could be compared to driving a car without a dashboard. Analytics give you the opportunity to steer your business. They help you understand what is working and what is not. Analytics are also the only way you will know if your advertising strategy delivers. Focus on collecting the analytics that will help in making decisions and improving your app advertising and marketing.

Mobile analytics empowers you to measure the performance and value of all your campaigns by each channel. To optimize the ROI of marketing spending, you can identify which channels deliver the best users based on engagement, monetization, and more.

There are a lot of companies to choose from, including Localytics, Apsalar, and Google. And their ranks will grow as the land grab continues and companies release new products or services for analytics. Mobile advertising networks want your business, and your app users' eyeballs are your "product." So take the time to identify what advertising is going to perform best with your app and for your user base to maximize your revenue. And measure your results!

Sections of this chapter were contributed by Asif Khan, LBMA; Itay Godot, Inneractive; James Coops, MobyAffiliates; Matt Lutz, AppClover; Ryan Morel, PlacePlay; and Gary Schwartz.

CHAPTER 14

Getting Discovered Across All Channels

You need to give your customers what they want. But it's a tough task if they don't know what they are looking for—or worse yet, have no idea your app exists. Also, if they have to navigate through multiple menus and sift through search boxes to find it, then you will likely lose them. Research shows that if app makers want users to download their apps, they need to remove the pain from the discovery process. But you also need to make sure you promote across all channels to spread the word that you have an amazing app!

It's Retail 101 All Over Again

If you ask any successful brick–and-mortar business owner, they'll most likely consider their list or database of customers and prospective customers their number one asset. And the reasons for this are simple: Loyal customers keep giving you their money; they become unpaid salespeople and advocates by spreading the word; and they don't cost you much, if anything, for repeat business, as opposed to the cost incurred on new customer acquisitions.

ESSENTIAL

In an interview *Appreneur Magazine* conducted with AppBusinessBrokers.com, founders Eric Owens and Mike Kemski stressed the importance of building a list from a company or app valuation standpoint. They shared that the bigger your e-mail list, the more your app or app business is worth.

Every marketing strategy you implement should funnel people back to your website—whether it's a website for your app or your app company's site. You want to build your online database of users (i.e., get their e-mail address). This is important for marketing your first app—and every app that follows.

Start with a List

Having an updated list of your customers is key to your success. Take a hypothetical situation: What happens if Apple decides it no longer likes your app and boots you from the store? Presto, your database of users is gone! You can no longer communicate with them, send them notifications, or generate revenue from them. And you just lost your biggest bargaining chip if you ever wanted to sell your app business. By creating your e-mail database via your app's website, you're building a safety net and asset for your app business at the same time.

Whatever marketing strategies you're implementing, whether it's social media, video marketing, e-mail marketing, guest blogging, sending smoke signals, or putting little rolled-up messages in bottles and sending them out to sea, make certain your end goal is to drive them back to your website and get them to op in to your database one way or another. And now that you know why, it's time to tackle the *how*. Here are some ways you can create engagement and funnel people back to your website and, ultimately, onto your e-mail list database.

First Impressions Are Key

The fact is, your app needs a place to hang its hat. And you need a place where you can focus all of your marketing efforts on driving people to your app and make all of your announcements regarding it or your future apps. And your community needs a place to learn about and interact and engage with you and your app.

When you think about the visual presentation of your website, promo, or demo videos, and your actual app, you should imagine you are presenting a five-star gourmet meal. The reality is, it doesn't matter how deliciously fun, addictive, or useful your website or app is; your users will ultimately make up their minds within the first couple of seconds of seeing it.

So you can either slide the equivalent of a can of spam in front of them or present them with a beautifully prepared, medium-rare Australian venison medallion proudly resting on a bed of herb-infused gnocchi and drizzled with a port wine reduction. Which one do you think they'll likely be more excited to consume—and, better yet, come back for seconds?

ESSENTIAL

Operators that maintain their own app stores or "store within a store" concepts can help you promote your app by offering you prominent placement. They can also help promote apps through ad campaigns, and through marketing arrangements. BlueVia (*www.bluevia.com*), a part of the global mobile operator Télefonica, sponsors the Apps Blog on the *Guardian* newspaper website in the United Kingdom and uses this as a promotion vehicle for developers who use the BlueVia platform and APIs in their apps. BlueVia was one of the first to offer developers APIs and is unique in offering app developers a revenue-share arrangement with developers on the use of messaging APIs.

You only get one chance to make that first impression, so you need to do what you can to create a meaningful and enjoyable experience for your users. A good first impression will dramatically improve your conversion rate—that is, how many users you convince to take action (such as joining your list, making an in-app purchase, upgrading from the freemium version of your app to the paid, etc.).

Resources

For some help creating an awesome-looking website, check out the Free App Wordpress themes on Apptamin (*www.apptamin.com*). To look at other App websites for inspiration, you can visit AppSites.com (*www.AppSites .com*). To put your best foot forward on your app's graphics, there are some great articles from some of the industry leaders on everything from icon design to app design principles found on AppClover.com (*www.appclover .com*). And to find great designers for your graphics, check out crowdspring. com (*www.crowdspring.com*), 99designs.com (*www.99designs.com*), dribbble.com, (*www.dribbble.com*), or elance.com (*www.elance.com*).

Social Proof

One additional, overlooked marketing tactic that you should utilize on your app's website, the app store listing, and the app itself, is "social proof." The more rave reviews of your app, the more followers of your website, Facebook, or Twitter profiles, the more likely it is that people will be intrigued and ultimately join in on the fun. It's human nature to favor the herd mentality. Think about it; have you ever visited a website that displayed thousands of Facebook fans or gone to a Twitter profile that had thousands of followers or checked out an app that had thousands of positive reviews? You probably thought, "Hey, if they all like it, I will too!"

The Discovery Dilemma

With over a million total apps available in Apple and Google app stores combined, and hundreds of thousands on the other platforms, the competition to get on consumers' handsets is fierce. It's becoming a shark-infested ocean, as hundreds of apps are added each and every day. App discovery remains, essentially, the unsolved challenge, and it's only getting worse. App quality, originality, innovation, and any other attempt to differentiate runs a very high risk of going unnoticed, with developers getting caught up in the "long tail" of apps that never achieve great popularity.

Jeff Bacon, director of Mobile Strategy at bitHeads (*www.bitheads.com*), a Canadian software house with clients including the *New York Times* and ESPN, sums it up best when he observes (talking about the Apple App Store) that "you're a fish in a very large pond."

Why Search Is Broken

Smartphones capable of delivering rich mobile Internet services and the proliferation of thousands of apps in over 125 app stores are creating increased demand for mobile search services. However, universal search is not suited for finding mobile apps. This is one huge reason why Apple acquired Chomp. Chomp, which uses semantics, artificial intelligence, and machine learning, vastly improves app search, allowing users to input more descriptive terms, such as "puzzle games," "kids games," and "expense trackers" to find (and discover) apps that they would otherwise not know existed.

Traditional universal search is one-size-fits-all. Whether you are a student, a scientist, or a silver surfer, Internet search engines deliver a similar set of results, regardless of your individual information needs. What's more, this way of searching, which is based on keywords and designed from the ground up to deliver results based on algorithms such as PageRank, is fundamentally flawed.

The burden is on your customer to have a pretty good idea of the app they want first; otherwise, it's tough to find it. To complicate matters, reams of research confirm that people seldom take the time or effort to look beyond the first ten results they receive. On mobile, a fiercely personal device with form factor issues (screen size, keypad, etc.), many people won't even go beyond the first five results.

For consumers, there are simply too many apps and far too much fragmentation in the actual app categories. They don't only need to know the name of your app, they need to know how the app store provider will most likely categorize it. To understand the confusion, imagine your app user is a customer walking into a grocery store needing only a few items and finding that all aisles and category labels have been eliminated and every product has been thrown into a pile on the floor.

For app developers, the situation is far worse. There are hundreds of thousands of apps, and most smartphone owners are only exposed to a handful. This is a nightmare scenario for developers, whose success depends on getting their apps "found" by consumers. "How will my apps be discovered?" is the number one question in the mind of app developers.

ESSENTIAL

The major app stores offer search capabilities. However, these app stores are commercially driven and search results are highly influenced by what the store wants the consumer to see, rather than being solely focused on producing relevant output. What's more, they are structured differently, so even if a user knows your app, she or he might not succeed in finding it, since it is not where one would logically expect to find it in the first place.

Search Solutions

Search is riddled with shortcomings and discovery is an ongoing dilemma. Sensing a business opportunity clever search companies such as Siri (licensed by Apple for voice-enabled mobile search), Goby (a combination location search and recommendation engine acquired by location-based services giant TeleNav), and Zite (a personalized news app for the iPad that automatically learns what people like in order to present them with more of the same category news and information; recently bought by CNN) are advancing to potentially fill the gap.

App search may currently be dominated by classic keyword search and "find similar" solutions that rely on social graphs and statistics, but this is changing—fast. Several new players, including Xyologic, Mobilewalla, and AppZapp, have new approaches that are gaining traction. Overall, mobile search is evolving as powerful enablers including semantics and artificial intelligence (AI) pave the way for a new generation of search, one that is far more precise and able to incorporate recommendations, human input, and context. This will accelerate as more apps become available and people become more sophisticated in their use of mobile devices and tablets to find and discover content, services, and—of course—apps.

Get Your Industry Street Cred

Making a name for yourself in the mobile industry is an important step in your app creation process. Just like with any business endeavor, it's all about networking and making connections. Here are a number of ways you can go about doing this.

Submit to Blogs

Write some great posts about the industry and submit them to other blogs, industry websites, your own website, or article directories, and find subtle ways to incorporate, or draw a connection to, your app. Do not write glorified advertisements for your app. This is a no-no. Whether you create an article that is a list of apps, like "The Top Fifteen Most Useful Sea Monkey Breeding Apps," or "Must-Have Full-Contact Origami Apps," the idea is to subtly work your app into the mix. Then, put your social media networks to work, add a post blurb on Facebook, and tweet about it. Get the word out and share the article with as many people as possible.

Forums

Also identify and frequent developer forums that match your interests and your app. Forums, by definition, are just online meeting places for like-minded folks with shared interests. What this really means to you is that there are highly targeted prospects just waiting for you to come along and spark up and cultivate a relationship with them.

Now, mind you, there is one golden rule to consider for this to work effectively. No direct marketing. Period. It's considered the equivalent of spam—it's another no-no, and you won't make any new friends by doing it. Unless someone directly asks you about your app, game, or website, don't mention it. Keep your conversations plug-free, as pushy marketing has become a thing of the past. If you contribute, and come from a genuine and authentic place, they'll seek you out.

Become a Public Authority

This is extremely effective and free. All you need to do is write and contribute articles around industry topics to relevant news sites, developer's blogs, and magazines. This delivers several benefits that money can't buy.

It extends you and your "brand" message to other audiences that you previously didn't have access to, it boosts your credibility and clout, and it is a little SEO (search engine optimization) "link juice" to your website and app in the process, ensuring once again that people out there will discover both.

The idea of just writing and submitting a bunch of articles may seem easy enough; however, there are a couple of prerequisites. You absolutely must have competency in what you are writing about, and you must also have a commitment to helping others. This isn't just a self-serving tactic. You have to deliver value.

ESSENTIAL

One outlet for your thought leadership in the app space is *Appreneur Magazine*, which has built an entire contributor-based website that's comprised of free content from some of the world's most successful app developers. The value exchange: Contributors like you provide the community there with tips around valuable app marketing and monetization strategies, and the site provides you with a platform for your ideas and a steady flow of prospects for your app.

What Are Your Options?

Bigger brands will be able to leverage their existing marketing channels to let customers know that they now have an app, which is fine, but this isn't an option for most developers. Some developers will turn to marketing and PR agencies, which will often tell them that they have to spend thousands on pay-per-click advertising, banner ad campaigns, SEO, and social media campaigns to have a chance of gaining awareness, and downloads, of their apps.

While undoubtedly effective, this is a prohibitively costly option for many, especially after taking into account the amount that will have been spent on development in the first place. One of the most effective ways to create buzz is through getting the app reviewed by trusted media.

This tried-and-tested approach is able to drive awareness about an app's release and can provide third-party endorsement from a trusted source. One single review won't always lead to a runaway success, but it does help to provide a few trusted and positive reviews at the top of a developer's app store copy that can give it that movie poster feel (you know—when movie posters put lots of great reviews on!).

This presents a further complication though. Journalists covering the app market can receive hundreds of requests or press releases per day. For premium apps, journalists have to spend time tracking down the developers, requesting codes to trial the app, and getting the information they need to write a review. These reviewers then spend anywhere between two minutes and a couple of hours on average testing it out and using it in a real-world situation. With so many apps being launched, reviewers can only get through so many apps a day.

As with getting an app onto the various stores, developers can't just expect the public and journalists to stumble across it. A certain amount of planning is required to ensure that the app has its best chance of discovery and success. One way of doing this is to use a service like appromoter. The service was created by a team of marketing and PR experts with a history of helping developers launch the latest apps to market. Appromoter works by showcasing all the latest apps, along with the information that journalists and bloggers need to write their reviews.

The service works by intelligently filtering through apps that get submitted and allowing journalists to discover the apps that are most relevant to them. For developers, it provides a cost-effective way to make an app stand out from the crowd. Appromoter gives each app a microsite with a dedicated URL and places all of the most important assets within a journalist's reach for them to download, including a promo code distribution system that enables journalists to easily review apps before their release, thereby giving them more time to review other apps or complete other pressing tasks.

It's also free for developers to upload their apps and collateral via appromoter. The collateral is a vital part of promoting an app and getting awareness out there. A common complaint among journalists and bloggers is the lack of material to support a new app launch. Many developers underestimate the amount of skill and talent that goes into writing an app description and press release and producing a video that portrays the benefits of an app to the media and the general public. For this reason, the team at appromoter offer a range of additional services that includes the all-important app description, press release writing and distribution, placement on featured listing websites and newsletters, SEO, App Store optimization, video production, localization and translation services, and advertising packages for mobile games.

Promoting Your App

Now that we've nailed down some of the fundamentals of marketing research, community engagement, and list building, it's time to start shifting to the tactical side of promoting your app. While apps are kind of like snowflakes, in that each one is unique, there are some proven foundational marketing strategies that work for all kinds of apps—from utility apps to games—that have stood the test of time.

Use Your App

A quick way to begin engaging your customer and building upon their trust in you is to simply send them a quick automated "thank you" for downloading and trying out the app. These two little words let your user know that you care and appreciate the fact that they downloaded the app. Simple, but powerful.

Sending notifications of upcoming features can be another way to keep your users engaged and anticipating something new from your app—new levels, functions, or features.

For game apps, one of the most powerful in-app engagement strategies is to introduce unlockable content. This gives the user the urge to continue playing until they unlock and can access everything. This is a tactic that you encounter every single day, and may not even realize it. It's called "the loop." And once a loop is open, human beings are hardwired to want to know how it ends. Whether it's your local news anchor giving a snippet of the night's juiciest story, which won't air until the end of the segment, or the QVC picture-in-picture showing us what's up next, or your favorite sitcom's "to be continued" cliff hanger, you *have* to know how it ends. And the same principle applies to apps.

Get Your Customers Involved

Setting up milestone achievements is yet another way to keep your users engaged. This is one of the principles of gaming theory and plays off of the competitive nature of humans. And while this is easy to understand when it comes to any game app—clear a level, earn points or rewards—it can also be applied to anything. Think about adding a referral system of some sort in which, upon referring a friend, you gain additional access, functions, or features. Think of the Dropbox model—when you refer someone else, you get more free space. Beats the alternative of paying for it.

Another way to promote your app is to have your users do it for you. You can integrate a "Tell-a-Friend" viral feature right into your app, which will encourage them to share it with family and friends they think might also enjoy it. But keep in mind, whenever you're asking someone to do something for you, the path of least resistance is to do 90 percent of the work for them. Make it as easy as possible by creating the prepopulated e-mail in which all they have to do is click "send." Then, their friends get an e-mail from them with a message and a link to check out your app, and presto, a new (free!) lead.

One last in-app marketing engagement strategy to continue nurturing your user's loyalty is to survey for feedback and ask them to rate your app (this helps with social proof too). Find out what your users love about your app so you can craft your marketing message around these points, and find out what they hate so you can create an update that betters the user experience, thus increasing future retention rates. A great feedback tool for this that integrates right into your app is called Apptentive.com (*www.apptentive.com*).

When you take this feedback and incorporate it into your new version, you'll be rewarded, since new versions move back to the top of what's new in your category. And since your app is new and improved, coupled with maintaining visibility, this can increase or at least maintain a steady flow of new users. If you're always working hard to make your app better, you'll consistently be rewarded with increased exposure, just because you're listening to your community and giving them what they want.

Creating Hype Through Reviews

Apple's saying "There's an app for that" is a bit of an understatement. The reality is, regardless of the category, there are probably thousands and thousands of apps for that. But the hard part for your potential customer is finding the one they want that's well suited for their needs—for instance, your new app. And that's the value of app review sites for developers. To get reviewed and featured on these sites means a lot more eyeballs on your app, and a lot more downloads.

The Life of an App Reviewer

There are a ton of great app review sites, such as AppReviewPros.com (*www.appreviewpros.com*), 148apps.com (*www.148apps.com*), iUseThis .com (*www.iusethis.com*), TUAW (The Unofficial Apple Weblog, *www.tuaw .com*), and AppSafari.com (*www.appsafari.com*), but they're not the only game in town. App reviewers vary from the salaried employees at the big tech sites to the independent entrepreneurs, small bloggers, columnists in industry publications like *Appreneur Magazine*, and those who write reviews as more of a hobby. And they all have different fee structures and practices. Some are free, some . . . not so much. Your job is to try a few out and find out which ones yield you the best results.

ESSENTIAL

Don't be afraid to reach out to bloggers for your niche market in other markets like Europe and Asia. And remember to include complimentary promo codes where necessary in your correspondence to them. Global outreach exposes your app to new audiences and goes a long way toward boosting discovery of your app.

Make It Visual

Another review avenue to head down is YouTube. Simply create and upload a video of your app (remember to optimize your keywords here too, as YouTube is the second-largest search engine around). Then, hit up some of the big YouTube app review channels, like UniqueApps, AppStoreReviewer, Appolicious, or CrazyMikesApps, to see if they'll review your app.

It might cost you a few bucks, but if you take into consideration the size of their channel's subscriber base, it might be worth shelling out for it.

Old-Fashioned Press Release

If you're looking to turn your app into big news, writing an attention-grabbing press release is key. And with so many developers seeking the media's attention these days, it's important to make your press release really pop and stand out from the crowd in order for it to get noticed by industry websites, reviewers, and bloggers. But don't get scared; writing a good press release doesn't require years of public relations experience or a master's degree in marketing. If you follow these simple guidelines, you'll be golden:

- **Make it newsworthy.** Only send press releases when you have actual news to share: the release of your app, major updates, new features, reaching noteworthy milestones (i.e., number of users or sales), being featured by Apple, and so on.
- **Follow the format of press releases.** Don't get creative here; just play by the rules. If you want to look like you know what you're doing (and be received better by the press) a quick trick is to seek out other published press releases for other apps and model what they did.
- **Avoid marketing jargon.** Your press release is not an advertisement, so stick to the facts.
- **Focus on the headline, summary, and opening sentence.** Much like your app in the app store, your press release is fighting for attention with many others. So by spending extra time on the opening hooks, you're less likely to be skipped over. Remember, this isn't a novel or story, no need to save the good stuff for the end, put those juicy facts where they belong, at the beginning to grab attention.

ESSENTIAL

To boost discovery of your app you can also hook up with other app developers and do a cross-promotion. In this scenario you promote each other's apps to your respective audiences via your apps and websites. Sharing your user bases with one another is a great win-win advertising tactic. Oh, and most of the time, if your user bases are similar in size, it's free.

Rethinking App Discovery

With roughly 15,000 new apps coming on line each week across the Apple, Android, BlackBerry, and Windows platforms, it's a wonder users can find apps at all. But why should users go to the app store to look for apps? Why not have the apps look for the users? This is the logic that has guided YouAPPi (*www.youappi.com*) and its business model, which is all about providing the opportunity for users to discover the apps they want, matched to their context, and not make them look in a centralized location such as an app store.

Apps by Recommendation: A Case Study

To enable a new approach to app discovery, while offering all the stakeholders in the app ecosystem a chance to generate additional revenue, YouAPPi has launched a cross-platform app recommendation and distribution solution.

In a nutshell, YouAPPi's smartAPP technology analyzes the content users are presently engaged with, the apps they have previously downloaded, and their preferences in order to match their interests with the apps they're most likely to want.

For app developers, which can include publishers, marketers, and other content providers, YouAPPi's app recommendation and distribution solution provides a marketing channel with a low cost-per-acquisition rate. The company's smartCONTROL interface provides a real-time dashboard offering an analysis of user behaviors, impressions, and installs.

The end goal is all about matching the users to the right apps. Think of it like dating. It's all about finding the right match. It's not about whether the guy or girl (or app) is beautiful or popular (ranked high in the app store). It's about finding the right match for each individual.

In order to prove that matching will get people exploring and downloading apps they might not have known about otherwise, YouAPPi uploaded its recommendation app to the iTunes store. Among the 12,100 users who downloaded this app, 3,630 users, or 30 percent, went on to download the app that was recommended to them by YouAPPi's technology. Put another way, YouAPPi played matchmaker between people and apps, promoting apps to different people based on the data it had collected and the app use it observed.

YouAPPi found that users respond to app recommendations. In fact, this matchmaking encouraged users to download more than 3,600 apps they might not have even known existed. YouAPPi also continued to measure app interaction among this sample to check if users went back to use these recommended apps often. The results show these users activated the app at least ten times a week. The discovery dilemma isn't entirely solved, but this example shows it might just be better to take the other way around. Don't wait for users to find your app; let the apps find them.

Moving to Automatic Discovery

With an avalanche of apps being offered, users can't say they don't have choices. But they can complain about the tedious process they have to endure to find apps they like. Against this backdrop, the vast majority of apps may as well not be offered at all! They are often buried in confusing, oftentimes counterintuitive, hierarchical menus and categories. Mobile devices, with their screen-size limitations and restricted input capabilities, only exacerbate the problem.

The solution to the discovery dilemma is being found in technologies that collect the clues people leave, such as their app download history, their preferences, and their context, to show and suggest apps that are relevant, and therefore valuable, to users on a more individual basis.

The race is on to supercharge app discovery by taking search out of the equation altogether. Companies like Hoopz are building a mix of capabilities that allow them to pursue a strategy of "zero-search," which will serve up apps (and other content) to users based on their context, possibly eliminating the need for a search box forever. It's a superior model because it is built from the ground up to deliver users more precise (hence relevant) suggestions and recommendations, in some cases even *before* users explicitly request them.

This revolution in the App Economy will start, as all such events do, with tiny steps. Hoopz suggests mobile app developers take matters (in this case, metadata) into their own hands. In practice, this means describing and tagging apps with every possible detail, including keywords, phrases, locations, differentiation from other applications, features, ratings, testimonials, and target profile. Why? Because while devices will likely pack more computing power, they will keep roughly the same size displays. Screens are not suited for the number of results from app store search queries (and neither is your customer's patience). Additionally, the huge growth in data, the failure of filters, and the lack of machine-readable metadata will persist.

The latter issue cannot be overstated. The Internet was created without a structure for machine-readable metadata, which is essential for high-precision and accurate search. In the meantime, text classification—tagging—is helping to create metadata. Looking ahead, new solutions and approaches will certainly emerge to address this problem—and they will create new services as well as new opportunities.

FACT

Developer Economics 2012 indicates that Facebook is far and away the most popular promotion channel employed by app developers. Of the 1,500+ developers surveyed, nearly half (47 percent) said they used a Facebook presence to drive discovery of their apps. Facebook claims to have sent over 160 million visitors to mobile app pages in March 2012 alone.

Moving forward, Hoopz even suggests that app developers should come together in communities and start grouping their applications according to their own categories, rather than leaving this task to the app stores, which have shown that this is their shortcoming, not strength.

Think it through and it would be possible for each app in a category to become a kind of lead generation tool for the other applications within the same category based on user choice and context. It's the Amazon model all over again—but much more potent.

Sections of this chapter were contributed by Akash Sureka, Hoopz; Ed Vause, Appromoter; Matt Lutz, AppClover; Moshe Vaknin, YouAPPi; and Dan Appelquist, BlueVia.

CHAPTER 15

Achieve Lasting Impact and Loyalty

Smart developers (that means you!) understand that selling apps is a serious business. But figuring out where to offer and how to monetize your app is just half the battle. You also need to work out a strategy to increase engagement and identify (and keep) your most loyal users. The build-it-and-they-will-come approach that marked the early days of the App Economy has been replaced by the hard truth that customer retention is what will distinguish app leaders from the also-rans.

Encouraging Lasting App Engagement

How effective you are at reaching and encouraging your customers to make your app part of their regular routine will make or break your app business. Winning is all about making the right choices to delight your target audience again and again. If the task sounds daunting, you are not alone.

The *Developer Economics 2012* survey found that developers everywhere struggle with ways to keep their existing customers engaged and interacting with the app over time. This is a must, since acquiring new customers is always more expensive than retaining existing ones.

Getting the approach right is ever more critical if you have chosen a freemium model to monetize your app. Your app business—making money via in-app purchases—rises or falls on how well you can keep customers coming back to purchase more features, levels, or other functionality they appreciate.

The Power of Push

What you make out of your business depends on the choices you make. Map out a comprehensive strategy that includes a toolbox of tactics to engage with customers every step of the journey. This is where the humble text message makes a big comeback.

FACT

Text messaging marked its twentieth anniversary last year (2012). Today, SMS can reach over 5.4 billion people around the world (over 77 percent of the world's population).

Messaging Is Direct Marketing

Text messaging is ubiquitous, universal, and widely regarded as the truly native language of mobile. It's the simplicity, pervasiveness, and sheer dominance of text messaging that has made it the world's leading data communication tool. People everywhere can (and do) use their mobile phones to send and receive text messages. Messaging enables instant self-expression and communication. It's also a form of direct and personal communication that allows you to get closer to your customer than any other communications channel ever invented.

Indeed, the vast majority of people open and read text messages within minutes of receiving them. But people aren't only reading messages; they are answering a call to action. Internal research from Optism, a mobile marketing solutions provider, reveals the vast majority (59 percent) of people respond to text messages within the first hour of receiving them. Moreover, 41 percent of people respond within just ten minutes of receiving the communication.

Connect and Communicate

Text messaging can extend the life of your app. This is the key takeaway of *Achieving App Impact: Using SMS to Encourage Interaction, Drive Loyalty*, a new white paper produced by MobileGroove and sponsored by tyntec. This white paper shows how developers can harness text messaging to encourage app users to try new features or explore other related apps. App developers can also re-activate users who have not engaged with their app in a while, or deleted it altogether.

ESSENTIAL

Push notifications allow an app to notify the user of new messages or events—even when the app is closed on the device. It's similar to a text message, but it doesn't offer global reach because it only serves smartphone owners. It's a great way for apps to interact with the smartphone in the background, whether it's a game app notifying the user of an event occurring in the multiplayer game world, or simply a mail application beeping as a new message appears in the user's inbox.

Text messaging can drive positive results for your app in a variety of scenarios:

- Build a closer relationship with your users. Earn their trust by talking and listening in a conversation delivered via SMS. Gain their permission and deliver messages they will appreciate about the apps they love.

- Drive additional app sales by introducing an "invite a friend" program powered by SMS. Empower your app users to be your app advocates by allowing them to spread the word. Provide an easy way for them to encourage friends to download your app and take it for a spin!
- Reach out to users who only open your app occasionally, and re-engage with users who have deleted your app altogether.

FACT

Veteran mobile author and analyst Tomi Ahonen estimates people sent 6.1 trillion text messages in 2011, up from 1.8 trillion in 2007. Meanwhile, research firm Informa Telecoms and Media reckons total SMS traffic will reach 8.7 trillion by 2015, up from over 5 trillion messages in 2010.

Push technology, in which users request that specific data be automatically sent to their computer or mobile device, has been around for nearly two decades. Now app makers are harnessing it to reach and engage their customers.

It's a vertical explosion, as app makers use push messaging to segment their audience, connect with their best users, reward customer interaction, send coupons and offers, trigger calls to action, and increase the overall time spent with their apps.

Of course, how you use text messaging or push messaging depends on your app, industry, and business objectives. Choose what works for you. But here are some helpful examples to start you brainstorming:

- **Entertainment and games:** Use push or text to deliver messages that motivate fans to keep playing the game, and—in doing so—generate the buzz around your app. Deliver bonus content and teasers and drive users to other media where you have a presence (like mobile websites or video/TV shows).
- **Social networking:** Whether it's fans of a band or cat lovers everywhere on the planet, your app must meet the communication needs of the

community. Use messaging to keep them up-to-date with new content and posts from members they follow; entice closer interactions; kickstart chat conversations; or just remind them of dates, events, and fun stuff to do.

- **Publishing:** Whether you have a popular blog or a local newspaper, you can harness messaging and notifications to deliver local alerts about what's happening nearby, promote issue/website highlights, encourage participation in a reader sweepstakes, and move your audience from reading single editions of your content to signing up for subscriptions.
- **Retail:** From unique boutiques to chain stores, use text messages and notifications to drive results deeper down in the purchase funnel. Invite your audience to rank and rate products you offer, deliver them information on product availability and new styles on offer, confirm shipping information and dates, send coupons, and enhance customer service.

But you don't need to "broadcast" your messages; you can also "get personal" and send them to a smaller segment of your audience. *You* decide and divide your users by such factors as demographics, postal code, or just the fact that they haven't used your app recently.

Extend the Life of Your App

Push notifications are what Urban Airship, a leading provider of push notifications, calls the "voice your app has with a consumer, when they are not actively engaged with your app." Push is also a win-win because it allows you a direct channel to deliver relevant information around your app and get even closer to your best customers.

Push notifications are the peanut butter to mobile apps' chocolate, making apps more sticky and delicious by not requiring them to be opened in order to provide value to users. These opt-in messages arrive on the home screen of mobile devices and can be customized with unique sounds and multimedia including video and form fields. And they come in a variety of flavors—customer service alerts, social updates, new content availability notices, relevant and requested offers, breaking news alerts, weather alerts, traffic alerts, reminders, and location-specific alerts.

ESSENTIAL

Today's consumer is finding out about new products, styles, and trends from friends, friends of friends, and perfect strangers. Now, more than ever, app developers need to "incentivize" app sharing and encourage recommendations.

Keep Your App Fresh

Push notifications are becoming a necessary element in a strategy aimed at increasing app engagement. The results are impressive—particularly if the messages are opt-in. In this scenario, the customers give their permission, showing they are genuinely interested in receiving messages from app makers that keep them posted on updates, breaking news, location-based opportunities, and other information the app maker wants to deliver.

In fact, the new Urban Airship Push Messaging Index, an analysis of the impact that push notifications have on mobile app users based on real-time data collected from the top push notification senders, found that using push notifications more than doubled customer retention over a six-month period after downloading an app. The index also measured the level of personal engagement a customer had with an application by measuring the total number of times an application was opened over intervals of one month, two months, and three months. Among the key findings: User engagement more than doubled for users receiving push notifications during the first month of download.

The bottom line: Use push messaging and follow good practice and you, too, can increase user engagement and build stronger relationships with your mobile app audience.

Good Push Drives Great Results

Initial studies have shown that push messaging can more than double both mobile app engagement and user retention. Adding location-aware capabilities—not just where consumers are right now but where they are over time—in addition to context such as direction of travel, can exponentially improve a brand's ability to deliver the right message at the right time.

However, this very personal real estate also demands a very targeted approach and one where the consumer is in control. Not all push messages are equally effective, and the personal nature of mobile devices demands that app makers treat this communications channel as a privilege.

Bad push is essentially no different than spam. This is why app developers, and other key players in the app ecosystem, need to marshal a movement toward good push so consumers have positive experiences.

The Push Bill of Rights

When done right, push notifications enable a brand to socialize with its best, connected, and engaged customers anytime, anywhere. Through its extensive interactions with over 65,000 push-enabled apps and more than 100 million consumers, Urban Airship (*www.urbanairship.com*), a push message service, has created a set of best practices it calls the Ten Rules of Good Push. Think of them as a kind of Bill of Rights for customers who have trusted you by opting in to your messages in the first place.

1. **Consider a customer's changing schedule:** Apps should have an easily accessible control panel where users can define a "quiet time" by adjusting a setting so that no messages are delivered between 10 P.M. and 7 A.M., for example. Another option would be to block out days on which they don't want to be bothered. Make this easy to find and easy to change.

2. **Engage your customers with relevant messages:** Every push notification should be data driven and triggered by what you already know about your customer. Consider what they have purchased, downloaded, or shared. Use location and any other bits of customer data and put your message in context. Think about tailoring push messages as if you were at a party. There might be limited situations where a shout out to the entire room is appropriate, but too much reliance on one-size-fits-all broadcast push notifications won't make the personal connections you need.

3. **Allow people to personalize their experience:** Give your customers a preference center to customize the content they want to see, and they will love you for it. The more exacting they can be in designing the push

content, the happier they will be to see it arrive. Burton Powder, the snowboard maker, does this really well. It allows app users to select the resorts, as well as day, time, and inches of snowfall that trigger a push notification. So does ESPN, which allows you to pick a sport, a team, what kind of alerts, and how often you'd like to get them.

4. **Stay consistent with your brand:** When you talk to customers through push notifications, do it the same way that you do on the web or in e-mail. Sit down and think about the best voice for your push notifications—and then execute.

5. **Deliver an engaging experience:** The goal is for users to look forward to your push communications. Sometimes just a glance as the message crosses a user's screen will deliver the value needed to build allegiance to your app. (For example, being the first to hear about breaking news or sports scores.) Once a user opens your message they should be transported into a rich media environment that is fresh, engaging, and entertaining.

6. **Make them better over time:** Ultimately, you can clearly tell which push notifications resonate best and which ones fall flat. At a minimum, you should know what your app open rates are and which messages drive higher or lower rates so that you can finely tune engagement.

7. **Adapt to the user's current situation:** Your push notifications should be smart enough to adapt to a user's current situation, including location. No one wants offers pushed to them with every step they take just because they are moving. Relevant apps consider everything they know about the user, including previous location information, in-app behaviors, stated preferences, and purchase history.

8. **Build trust before asking for push and/or location opt-in:** The default and incorrect approach is to ask users to opt in to push and location sharing as soon as they download and open an app. Instead you should allow them to explore the app and see all the features and content it offers, and then link the opt-in ask to an action—whether it is posting their first comment, creating a user account, turning on a specific feature, or beating the first level of a game. Even looking to delay the ask until the second, third, or fourth time they open the app will improve opt-in rates.

9. **Be specific about the value that push and location will offer:** Explain exactly why consumers should opt in for push messaging or sharing location within your app, being clear about the value that you will be

able to deliver to them. Have this explanation in welcome screens, privacy policies, or as a page where the standard Allow/Don't Allow Push Notifications pop-up appears.

10. **Serve your users' best interest now to get the sell later:** Mobile devices are tools for communications and to help people get things done. Don't interrupt all the time with advertising content. Users may not appreciate that. But if you create award-winning service experiences, they will remember and reward you in the long term. Next time you get ready to submit a mobile communication, whether it's a push message or an SMS, ask yourself, "Is this a service call or a sales call?"

With good push, you can be at the center of your user's life. But with bad push, you risk being perceived as an annoyance, or worse, an invasion of their personal space, which can cause them to turn off push or delete your app entirely.

FACT

Look for A2P (Application-to-Person) messaging to skyrocket. A key driver here is the rise of mobile apps and the realization by app developers that messaging is an excellent vehicle to deliver content, information, and updates to their users. According to Juniper Research, A2P messaging will soon overtake humble person-to-person SMS to be worth $70.1 billion by 2016.

Serve Your Loyal App Users

Since it is such a personal process, push notifications will allow you to connect with your best customers. But you can also reap huge business benefits if you market, promote, and connect with your most loyal users.

Quality, Not Quantity

Since the earliest days of mobile apps, the goal for most app developers has been to climb the ranks in the App Store or Google Play. The higher the rank, the more revenue their app would produce . . . or so the thinking went.

Against this backdrop, the pressure to deliver the download numbers was relentless. There was pressure to not only generate downloads in high volumes, but to do so quickly at extremely low cost per installs (CPIs). So app developers did what seemed only reasonable: They spent to reach their download goals and top chart positions, often overpaying mobile app ad networks. And they hoped that downloads would convert into users who engaged with their app and delivered a return.

App developers invested large sums of money with the ad networks to drive these high download volumes. And as developers watched their budgets dwindle, they pushed for lower-cost downloads. What app marketers were experiencing was analogous to the early days of online marketing. Early on, online marketers focused on metrics such as website visits, page views, and clicks. But as the industry matured, marketers learned that, though those were important metrics, it was really the end result metric that mattered most—revenue.

The lesson? Quantity looks good, but quality drives business.

And today, mobile app developers have arrived at the same conclusion. High download volumes may look good and feel right, but most downloads don't translate into conversions or revenue. Some developers may never launch their app at all, and some launches are one-time events that never translate into purchases, registrations, or repeat visits. In fact, users often abandon an app after a single use and the app experiences a high churn rate.

To monetize apps and build successful app businesses, marketers have concluded that loyal user engagement is more important than high download volume.

App Download Metrics vs. Quality, Loyal App User Metrics

App clicks, downloads, and launches can be misleading metrics. How many times do users return? Have they made a purchase? Did they register? Mobile app marketer Fiksu (*www.fiksu.com*) provides some answers based on real data that demonstrates how concentrating on generating high-quality downloads with loyal users can deliver an ROI much higher than a pure volume approach.

In this analysis, a six-dollar value was assigned to any quality, loyal user who registered within the app. The sample model compares the difference between emphasizing download volume versus focusing on loyal users. In the volume-focused campaign, spending $40,000 might generate 110,000 downloads for a cost-per-download of $0.36. The same $40,000 spent on a loyal user campaign might generate 27,500 downloads at a cost-per-download of $1.45.

The metrics in the volume-based download acquisition model look much better than the metrics in the loyal user acquisition model (download volume is four times higher, and the cost per download is substantially less). However, a closer look at the numbers reveals a different story.

Using the same numbers, the volume model actually only delivered a 5 percent conversion rate on registrations, yielding 5,500 conversions at a cost of $7.28 each. However, the loyal user model delivered closer to a 35 percent conversion rate, generating 9,625 registrations at a cost of $4.16. At the $6 value per registration, the volume-focused approach resulted in a negative ROI of $7,000, while the loyal user model netted an $18,000 gain.

This example shows that targeting high-quality, loyal users is profitable. The bottom line: A strategy based on driving high volumes of low-cost downloads actually puts the business in the red.

Evolution to Quality, Loyal User Acquisition

Clearly, the most effective way to build an app business is to have a user acquisition strategy based on ROI metrics rather than download metrics, as it enables you to truly scale your business in an ROI-positive way. It's no shock that app developers are moving from the volume-based download acquisition model to an ROI-based loyal user acquisition model. In this model, downloads may not be as high, but loyal usage is much higher, leading to more revenue with substantially higher marketing ROI.

To monetize an app, you have to develop a compelling app that people want to use. Just as importantly, you need to get the app into the hands of the right people—loyal users. What exactly constitutes a loyal user will vary from app to app, but simply put, a loyal user is one who takes an action that ties back to your ROI. In other words, it's someone who has made an in-app purchase, launched your app many times, reached a certain level in your game, or registered as a user.

Marketing Strategies for Acquiring Loyal App Users

Now that you know who they are, you need to map out a strategy to acquire more loyal users and keep them coming back. After all, they are essential if you want to build a thriving app business. Here are some proven tactics from mobile app marketers responsible for some of the world's leading app brands and publishers.

1. **Establish loyal user goals.** First, you need to define loyal user characteristics. Is it a repeat visitor who drives advertising revenue? Or a user who makes an in-app purchase? Or maybe it is a combination? Then you need to define the lifetime value of a loyal user to determine how much to spend to acquire one, and lastly, define cost per conversion target to understand how much to pay for a given traffic volume to deliver the best possible ROI.

2. **Promote the app across several traffic sources (incentive networks, non-incentive, and real-time bidding exchanges).** Successful app marketers have found that working with a large number of traffic sources can improve loyal user acquisition up to five times. By working with several traffic sources, you can:

   - Improve the likelihood of finding the traffic sources that deliver loyal users
   - Avoid subjecting apps to audience saturation
   - Maintain a more consistent stream of traffic in the face of inevitable traffic fluctuations
   - Sustain campaign performance over longer periods
   - Increase campaign ROI by identifying opportunities for lower acquisition costs

   Identifying high-performing traffic sources and exploiting those sources is a key driver to acquiring loyal users. And the best way to do this is to start with a large number of traffic sources.

3. **Track postdownload activity.** As ad campaigns run, use one of the available technologies to track postdownload activities and tie each of these activities back to the traffic source, ad creative, and other campaign

variables. Use this data to identify the traffic sources and campaigns that are driving loyal users.

4. **Optimize ad campaigns in real time.** With the insights gained from post-download activity tracking, optimize advertising spend in real time toward traffic sources that are delivering the users who take the actions you want. Eliminate inefficient ad spending and shift that money to the traffic sources and campaigns that are delivering loyal users.

5. **Employ a campaign optimization solution.** Optimizing campaigns based on postdownload actions is a complex task. The good news is that automated solutions are available that track postdownload actions and optimize ad campaigns in real time to the traffic sources that generate loyal users. Look into the available solutions and select the one that's right for you.

6. **Conduct ongoing closed-loop analysis.** Analysis of marketing campaign performance, postdownload usage, and revenue return will help you understand at a granular level what is truly driving your app business. Use this information to continuously refine your marketing campaigns.

7. **Finally, conduct a loyal user acquisition test.** Assign a portion of your marketing budget to loyal user acquisition and conduct a test that employs the strategies outlined above. Such a test will provide you the opportunity to learn how a strategy of loyal user acquisition can impact your bottom line.

To build a thriving app business, app marketers need to think beyond the download to a mindset of acquiring quality, loyal users. It is loyal users who deliver revenue return, turn into word-of-mouth advocates and become repeat customers.

Sections from this chapter were contributored by Scott Townsend, Urban Airship; Viki Zabala, Fiksu.

CHAPTER 16

Get Seriously Social

Your shiny new app is a glistening iceberg, and creating and publishing it is just the tip. The marketing work starts the moment you conceive the idea for your app, and it continues on throughout the entire life of your app as you build a new—or engage your existing—audience with the story of you and your app in the different online communities in which you participate and/or create.

Build Your Community

Having spent the last few weeks or months locked away brainstorming, designing, planning, coding, testing, and playing around with your app, it is now on the market and ready to make money for you. There's just one problem, though; you're not the only one vying for attention. On average, there are 500 apps launched on a daily basis across the major mobile operating systems. If you're not a major developer, brand, or publisher, you need a way to get your app to stand out from the crowd, and it may require a substantial investment to generate a bit of marketing hype to get the attention of potential customers.

QUESTION

What makes a "community"?
Everywhere your audience—your app's target and existing users—congregate online is a community. A new or existing user forum or online discussion group is a community. Your connections on each of your different social media networks channels are communities. The LinkedIn groups you create and/or participate in? They are all communities.

Many developers hope that once they get past the development and approval process then they'll magically become the next Angry Birds. In the quest to reach the top, some developers will invest in getting their app featured on the front page of the app stores' websites. In the United States, for example, large developers are now spending an average of over $100,000 to push their apps into the "Top Twenty" lists of Apple's iTunes and Android's Marketplace stores. For most developers, this just isn't possible.

It is crucial to focus on building communities of brand advocates—who may or may not ever be users—for your app. While you might be tempted to wait until the app has launched, you should start building your communities when you start creating your app. It's a process that you will need to manage, although you may also want to pay someone to maintain your social media presence for you. If this is what you are thinking, then just be sure that the app is financially viable before you invest in a social media or community manager.

Human Touch

Above all, no matter what option you choose when building your audience, be genuine and human in all your communications. Many app developers are nervous about managing Facebook or Twitter audiences because of their lack of formal marketing training; they often wonder if they have the communications skills and writing ability to handle these tasks themselves. Even if you're not a natural wordsmith, don't feel intimidated by the prospect of speaking to your fans and users via social media.

An app developer's lack of formal marketing experience can actually be a benefit: Social media success often goes to people who are good at "being real" rather than being a silver-tongued spokesperson. Instead of trying to pick the right words or trying to sound "bigger than you are," embrace the fact that you are a small developer. Be yourself. Share your passion for your app, be generous to your audience (by thanking them for following you and by promptly responding to their feedback, questions, and ideas), and make clear that your social media efforts are in service to create better apps for your customers, not "getting rich and famous" for its own sake.

ESSENTIAL

You need to have a human voice and respond to your users' questions and ideas promptly and authentically—but you don't have to allow social media to take up too much of your time every day.

There are a lot of great tools available to help you automate your social media activity. For example, Hootsuite (*www.hootsuite.com*) is a social media management tool that enables you to schedule your different social media posts in advance. If you get a bunch of great ideas for social media content, you can save them and space them out in Hootsuite so they publish on a regular basis. Tools like Hootsuite are a great way to maintain a consistent content publishing schedule, giving your users a regular flow of new content, while you maximize your time without having to stay logged into Twitter, Facebook, or LinkedIn all day every day.

If you are not going to actively monitor each of your different social networks, either manually or through the use of a social analytics tool, be sure to set up e-mail or text alerts for each network so you know in real time when you're being talked about and can respond in a timely and appropriate manner.

Here are ten helpful tips to help you get started on this exciting journey:

- **Define your objectives.** Why are you building an audience? Is it to get feedback on the app? Are you going to be launching further apps and want to be able to become a trusted developer? Are you going to be offering discounts or special offers that you want to communicate? Is it to provide support or networking? Before you start, you need to define your objectives so you can choose the best strategies to achieve them.
- **Know who your users are.** To create an online community of fans and users, you need to understand who they are. It's vital you reach out to and engage with your initial users so that you can understand what they expect from you, what they think about your app, and what value they want to receive from their participation in different online communities. Try putting something like Flurry Analytics in your app. It's completely free and gives you a very deep overview of who is using your app and what they're doing.
- **Tell your story.** Use the unique story of what led you to create your app to build your online presence. Communication on social media will often be the primary channel your users use to reach out directly to you, and for you and your app to speak to your new and existing users. Make sure your communications are humanized so that users know there's a real person who cares behind your social media accounts and not just a robot.
- **Take your time.** This is very important. Your community isn't going to multiply overnight. By focusing on engaging one person at a time, you will find your community will gain its own momentum. Be patient and be vigilant.

- **Add social networking to your product.** This is quite possibly the easiest way to build your audience; if you want people to share, make it easy. You can include links to Twitter, Facebook, and your app's group on LinkedIn in your app itself and invite people to connect with you for updates. You can then use this to give them opportunities to tell their friends about your app. If you have multiple apps in the market, it's always good to cross-promote between your apps as well. For game developers there are also some great community-building solutions out there such as Papaya or Openfeint.

- **Connect your members.** After gaining social connections, the work isn't over. Now it's about building a place where people in your audience can talk to each other. This will mean you won't have to always lead with the talking but can continue to listen and make the community self-sustaining. Make sure your online community finds value in their involvement—focus on building that value, and your community will not only stick around, but also become a huge supporter of your company and its products.

- **Give privileges.** The privileges for nonmembers and members should be different, thus providing an incentive to join and participate. For example, you might allow audience members access to special content, new information, discounted products, and so on.

- **Highlight influential members.** Influential audience members have a direct impact on the behavior of other users in the community. By engaging and highlighting your most "influential" members, you'll empower them to help you to further develop your audience.

QUESTION

How do you identify the most influential members in your different online communities, be they your connections on social networks, members of your LinkedIn groups, or members of app user/fan communities in which you participate?

Many social analytics tools, including Hootsuite, now include scores that measure people's level of social influence, and the topics in which they are most influential online. You can also look up people directly on social influence measurement sites Klout, Kred, and PeerIndex.

- **Track the impact of your posts.** Social is measurable; note which of your efforts get the best responses, and try to do more like them.

Social Media Crash Course

While social media can be great for communicating with your users, you will need to know the basics to get started. There are thousands of social networks, each one having its own "personality," culture, and etiquette. At the time of writing, the seven biggest social networks are: Facebook, LinkedIn, Twitter, Pinterest, Instagram, Google+, and YouTube.

What *Are* Social Networks Anyway?

Social networks are the web and mobile social media communication channels on which everyone can freely create and publish content. What makes social media so "disruptive" to traditional "command and control," *Mad Men*–era marketers? Social networks have eliminated the historically high costs, long cycle times, and need for specialized professionals that have been the hallmarks of traditional mass media.

Publishing your content via social media is now virtually free (yes, you still need to have a device and an Internet connection). Moreover, it is instantaneous.

Social media communications also have the opportunity to go "viral." That is, anyone with a connected device anywhere on the globe can use social media networks to spread content across various platforms. Just look at the role of social media in recent political uprisings across the Middle East and in the U.S. presidential elections. Turn this around and it means social media gives everyone the same potential to have a loud and influential voice. In the same vein, *you* (and your users) can have more social influence and more authority about a product or service than the companies that manufactured it.

Think about this for a moment. To you, the app maker, this is exhilarating. Based solely on your expertise, your value, your ability to communicate, and even your snarkiness, you can quickly become the most influential voice online (and off!) on any topic. You can wield social media to ensure your app makes its mark on the market.

But there is a catch: You can't sell it outright.

What trips up marketers, especially those who excel at the traditional "hard sell," is that you do not "sell" on social media the same way that you sell on your other channels. Posting "My app is great! Buy my app, now!" will fail to engage your audience and will likely get you flagged as a spammer.

Write this one down. To effectively use social media to sell and market your mobile app, you do not *sell* on social networks. Instead, you participate, you educate, you share, you inform, you geek out, you entertain, and you engage.

ESSENTIAL

Tips for Twitter success from "Fast Start 40: The Checklist to Help You Get the Most Out of Twitter to Market and Sell Your Mobile App" include using your business's keywords in your Twitter bio, tweeting at least three times a day, and adding your own comment or analysis when retweeting others' content.

You Need to Be You

Social networks are by definition personal, and they work best when you participate in them as the real, authentic, named you, vs. a brand or a logo. People connect emotionally with other people more readily than they do with a brand. So you point out that Nike has almost a million followers on Twitter. You're not Nike . . . yet. Generously sharing (not selling) your passion, your story, and your expertise, is the "secret sauce" of effectively using social media to effectively sell and market your mobile app.

QUESTION

What topics should you *avoid* talking about on social media? Regardless of the social network, think of everything you say as being instantaneous, permanent, and viral. Avoid the same polarizing topics you would at a cocktail party with people you don't know—sex, politics, and religion—unless, of course, you're developing a sex app, politics app, or religion app! Remember to be positive. Why? Because positive content engages!

Another significant characteristic of social media networks is that they are changing, constantly. By the time you read this, Facebook will have made yet another change impacting what you need to do on your brand page, the pundits will either be loving or hating Google+ (varies hourly), and there will be hundreds of new free, freemium, and for-fee tools to help you get the most out of your marketing efforts on each and every social network.

Which Social Networks Should You Use?

In addition to the top seven mainstream social networks, there are literally thousands of niche social networks. The obvious starting points are Twitter and Facebook for consumer apps and LinkedIn for business apps. Google+ is particularly useful for connecting Android fans, as Google runs many of its developer relations work through Google+. But Instagram, Pinterest, Quora, and other networks may work for you, too. If your app is focused on imagery and design, then Instagram and Pinterest may be your most valuable.

If you want to be seen as a thought leader on certain topics, then LinkedIn and Quora will be relevant. If you're short on time, then Twitter wins hands down. Also, every industry sector has its own networks and online groups, and there are many LinkedIn and online user forums/discussion groups with new ones emerging all the time. It really depends on the combination of where your customers and clients are likely to hang out and which of the networks you feel most comfortable with, and making these networks part of your daily sales and marketing routine.

Keep in mind that the most valuable social networks are the ones your target and existing customers use. For example, if your app is targeting pediatric oncology nurses, you will want to actively participate on the various nursing, pediatric, and medical oncology social networks. If, however, your app is targeting cat owners, you will want to participate on one of the many social networks dedicated to cat owners. Take the time to ask your target customers where they spend their time online.

Social Media Marketing Checklist

As you are working on building your social media presence, remember:

- **Fish where your fish are.** Identify the social networks most relevant to the audience(s) for your app.
- **You snooze, you lose.** If you have not yet already done so, begin participating on these networks *today*. Do not make the mistake of waiting until your app launches, or even waiting until you know exactly what your app is going to do. Actively participating on social media is now a requirement for launching a successful mobile app.
- **You are your app's most valuable marketing asset.** Be you. Be real. Be authentic. Be transparent. What does this mean in plain English? If you are an engineering geek who loves the Red Sox and Batman, geek out. Write about what moves you, not in marketing speak, but in your own words. Passion + content = success. Likewise, if you're a guy or gal who loves technology and has a great sense of humor, be funny and talk about technology. Marketers struggle constantly with differentiation. Fortunately, there is only one weird and wonderful you; use it to your app's advantage!
- **Be consistent.** While it's valuable to post often (note that recommended frequency varies by social network), it's more important to post consistently. A social media network is *not* a "set it and forget it" marketing effort; it is a fireplace. While it can take time to build it to a roaring fire, the fire will "go out" very quickly when neglected.

Social Media Skills Test

While you don't need to understand how a car engine works to drive a car, you do need to have a few basic skills to effectively use social media. You don't need to be an expert, of course. Just learn what you can at your own pace.

Pop Quiz

Take the following three-question skills test to see if your being on social media will do your mobile app more harm than good.

1. Can you write clearly and with correct grammar in the language(s) of the customers you are targeting?

2. Do you have a basic understanding of sales, marketing, PR, issue/crisis management, and customer service strategy?
3. Are you a consistently positive/upbeat person?

If you answered "no" to any of these questions, you should *not* be an active voice on social media. Just because you can drive a car doesn't mean you should enter the Indy 500. The good news in this is that communication via social media itself is an incredible tool for learning, no matter what your subject of interest. Be the thought leader in your space (even if your app is a restaurant tip calculator), provide the greatest value to your audience, dominate your industry, and crush your competitors!

Networking

To network or not to network? That is the question. You're probably thinking, "What on earth does networking have to do with a book about mobile applications and how to sell them?" Many think networking is just a waste of time, but networking, both on- and offline, has an important role to play in the range of tools available to you to market and sell your mobile apps.

There are many definitions of *networking*, and you can look them up yourself on the Internet, but it is simply the practice of making contact and exchanging information with other people (be they individuals, companies, groups, or communities) who have interests in similar areas to you. In making those relationships, the goal is to create mutually beneficial relationships. The people you are networking with could be potential app users, commercial partners, investors, advocates, and more.

So what might you get from networking? It is important to work out what you want to get out of networking rather than just leaping in, otherwise you will find that it consumes a lot of your time for very little reward. And it's very easily done. There may well be multiple reasons for networking, and these will change over time as you grow as a businessperson and app developer.

Some Potential Goals

User insight. There's nothing like getting feedback from real people who are your potential customer base. It is important to go beyond the comfort zone of your friends and family members, who will always say nice things about your app, to reach people who will be your potential customers and hear what they really think about your concept, your design, your content, or your business model. You don't have to incorporate every single thing they say, but getting that feedback early on will help you work out exactly what you should be doing and how to reach your target market segments.

1. **Raise your profile.** It's valuable to raise your profile online and offline. If you're making apps in a certain sector or vertical, then being seen to be knowledgeable in that area is going to help you by giving you additional credibility and gravitas. You can do this online by writing articles for your own site and/or blog and doing guest articles for others and using Twitter, Facebook, LinkedIn, Google+, and other platforms effectively to spread the word. In the offline world, being a spokesperson or go-to person for quotes and for speaking opportunities will help raise your profile and in turn, raise the profile of the app or apps you're creating and distributing.

2. **Get reviews.** Reviews are the lifeblood of all the app stores. Without them, no one will take you seriously. There's a whole other section about this in the book. Meeting people via networking will help you get more reviews and help you understand the people behind the reviews so you can respond to them accordingly when there is negative feedback.

3. **Find partners.** Partnerships are not a guarantee of success, but they can sure help when they work well together. Traditional media owners (newspapers, TV, radio, and more) are looking for new ways to monetize their audience. And where do these media owners find new partners? Very often at conferences and meet-ups.

4. **Get investments.** Investors, whether they are angel or VCs (venure capital), are all attending mobile meet-ups these days. Mobile is the new black and where the investment money is headed. If you don't have a mobile strategy, you're unlikely to get investment for a digital business. And guess what; a lot of those VCs hang out at networking events, conferences, hack days, and tech meet-ups. If you need investment, those are good places to start to meet the right people.

5. **Keep up-to-date.** When you're busy building your app, it can be hard to keep up with the latest updates with regard to mobile technology and

devices as well as the latest trends and tips in app marketing. Mobile meet-ups and networking events are great places to get a quick lowdown on what's new and what's hot.

6. **Stay connected.** A problem shared is a problem halved—so pursue a discussion with your peers. It's lonely setting up your own business, especially if you've been used to a team environment in an office. If you're a home worker, there will be days when you don't speak to anyone at all but have your head buried in code. It's important to realize that you do need other people and a problem shared can be a problem halved, as the old saying goes. Networking with peers is a really good way to share ideas and ask questions to replace those water-cooler moments in the office that you no longer have. It's a chance to create a peer network where you can help each other forge ahead with your app businesses. And it helps you realize that you are not alone in this game and that there are other people like you out there with problems and issues similar to yours. By pooling efforts, you can save time and save your sanity.

Stimulate the Brain Cells

Networking doesn't all have to be about mobile technology. Indeed, you should be going out to events and reading things that stimulate your creative side too. Apps aren't just about coding. There's a lot more to them than that, and they need more than a little creativity to get the edge. Going to events and reading things about topics you might not normally go for should show you something new and help you think differently.

It also helps to get out of the house sometimes and see real people and do some "networking." Taking a "creative break" for a movie, play, band, poetry jam, or art show will help clear your mind and give you a different perspective. And hey, you never know who you might meet there. Have your business cards on you just in case!

Time Is on Your Side

It's important to remember that you can't do everything right away, though you will probably want to. You have to start somewhere and keep going and keep growing at a level that's comfortable for you and no one else.

It's not always easy to find the energy and confidence to start networking. It doesn't always come naturally, and there are times when the last thing you want to do is go and be social with the world. That happens to the best of us, and that's fine. It does get easier if you're prepared and have given it some thought beforehand.

Know Your App

This may seem trite, but it's amazing how many really smart people are unable to explain what their app does in a clear and simple manner. Too often, the explanation is overly complicated and is full of technical or industry jargon—or worse still, acronyms—of which only they know the meaning. So practice telling people what you or your app does and then practice again. And then do it some more. Make sure that anyone can understand it, even your grandmother to the point that she could explain it to someone else.

Also make sure you have a quick demo at the ready to use so you can back up your pitch with some slides, screenshots, or whatever. Make sure you have a version of the demo that can work offline—there's no guarantee of decent connectivity at a networking venue. And keep it short. Really short! People are busier than ever, and you never know when you'll get a chance to deliver your one-minute pitch so make sure you keep it to a minute or less.

Tools of the Trade

You'll need business cards. Lots of them, and in a font size that is easy to read, preferably without the need of reading glasses. You're working in the app business; make sure you have the right devices with you, whether that's a phone, a tablet, or a laptop. And bring your charger too, just in case! If you're at a conference all day, you will run out of juice. It's also worth eating before you head out to a networking event, especially if it's an evening one. A rumbly tummy or getting squiffy on one too many glasses of wine doesn't give a great impression. Don't rely on there being canapés to fill you up.

Talk to People

This is harder than it sounds. You're faced with a room full of people all busy talking to each other and all looking much more important and knowledgeable than you. And it's daunting. And you want to turn around and simply leave them all to it. The best way to counteract this is to volunteer at an event. Volunteers are the backbone of a lot of networking events and conferences. If you volunteer your time, it means you have a role at the event and it gives you a good reason to talk to people. Typical roles might be helping with the technical setup, welcoming people and signing them in, helping at the bar, live tweeting the event, taking photographs, and lots more. This also gets you in with the organizers of the event who usually know a lot of people and may be able to make introductions for you. This approach is much more about making friends than formal business networking.

If you're just an attendee, that's fine too, but you need to make sure that you speak to a lot of people. Not everyone in the room will be useful or relevant to you, and there isn't enough time to speak to absolutely everyone, but networking is partly a numbers game. This means that you need to make the effort to "work the room." The idea here is that you don't go straight in with your pitch, but you allow the other person to lead and then you can work in your short pitch (and do make sure it's short!) by weaving it into the conversation.

When you speak to a lot of people, it's very easy to forget who they are, so make it a habit to write a few short notes about them on their business card or in your notebook. And then move on. Don't spend the entire event with just one person—well, not unless it's very much worth your while to do so. You are there to meet people, so it's perfectly acceptable to end one conversation and then move on to another. Or bring someone new into the conversation who might be standing on their own at a loose end or who looks a bit lost. If you've come away from an evening networking event with three good contacts to follow up on, you've done well.

Follow Up

This is really important. There is no point meeting lots of interesting people and then forgetting all about them, because, well, they'll probably forget you too. It is well worth while to keep a database of contacts. You may want to use something like Sage Act, Salesforce.com, Card Munch, or your own database. Be methodical about adding people to your database and noting key points about that person and what the followup action is. Then do the follow-up action. This is most likely to be an e-mail followup, or it could be that you tweet them later that evening or the next day.

This is the start of building a relationship with that person by putting yourself on their radar in a gentle way. This isn't about pitching relentlessly; it's about building up a rapport and a friendship that could be mutually beneficial. Networking is very much a two-way street, and you need to give first before you can reap the rewards. Equally, the rewards are unlikely to be linear or directly or immediately reciprocated. The connections are often looser and take longer to grow. They need nurturing over time. Ultimately, it's not about whom you know, it's about who knows you and what they know about you so they can recommend you, your service, or your app to other people.

Time Wasters

There are lots of time wasters out there. They're full of enthusiasm. They promise you the earth—well, introductions, cash, clients, whatever. And then they just don't deliver. Or worse still, they're energy vampires and just sap you of your energy. Over time, you'll begin to spot these people and you'll learn the signs. They can be subtle, and this just comes with time and experience and a dollop of cynicism. Very often, people can just get carried away and overpromise and under deliver. It's part of the human condition.

Sometimes, they are people who were never going to deliver but just wanted to be polite and wanted to be liked. Sometimes, they're just timewasters and are to be avoided. Sometimes people are just busy and forget what they promised, or something else took priority on their to-do list. It happens. Don't be offended; it's rarely personal. And don't be put off. As the old saying goes, you need to kiss a lot of frogs before you meet your prince.

So you're now prepared. You have your pitch ready. You have your business cards. Your Twitter profile is up and running and you have a demo slide deck.

Where Do You Find the Right People?

If you live in a big city, there will be a large selection of events and networking groups. These will vary from general business networking groups, where members meet very regularly and are committed to making introductions on behalf of each other, to more specialist tech-oriented or industry-sector meet-ups. Good places to find these are on LinkedIn, Meetup .com, and Facebook. And since you're in mobile communication, then you could do worse than look up your nearest Mobile Monday chapter (*www .MobileMonday.net)* and find out when their next meeting is. You could also look up relevant conferences you might be interested in attending and see what media and association partners they have. Those media outlets and associations will often have meet-ups or networking groups (on- or offline) as part of their remit.

In other cities, you'll find tech coworking spaces, and these are a great way to connect with like-minded people. Very often, they will also run or host networking or training events.

ESSENTIAL

The easier you can make it for your users to talk about your app and share their experiences with their friends and followers, the more likely you will be to attract extra eyeballs and get extra clicks, often with no extra effort or investment on your part.

Part of Your Routine

You're busy. You have a million and one things to do every day. And lots of things are competing for your attention. It's very easy to leave out the online networking and focus on things you find easier to do or figure are more important. Online networking is important. It's a direct communication channel with your customers and potential customers. It's your visible, constantly updated, online presence. It's how you are viewed by the world at large so it's important to maintain that presence.

There's no suggestion here of spending your whole day doing this, but making it a regular thing to check your various social networks and update them with the odd comment, article, or whatever, is as important as checking your download stats and your e-mail. You can manage this stuff from your mobile device or your laptop. Set aside a little time every day to do it and it will soon become part of your routine. The people you talk to will become your friends and your business colleagues, your mentors and your customers, and the conversations will inform your next business decisions.

Just remember that networking is about making friends and being human. It doesn't have to be hard. With a bit of simple preparation, it should be a fun thing to do and will enhance your business at the same time

Sections of this chapter were contributed by Helen Keegan, Founder, Heroes of the Mobile Fringe; Ken Herron, social marketer and cool hunter.

CHAPTER 17

App Maintenance

Congratulations! You've launched your app. Now what? All developers face many challenges after they have launched their app. Tracking bugs and errors is, by far, the most frequent postlaunch headache, as reported by 38 percent of developers. There is no direct feedback channel between users and developers, and no out-of-box means to monitor the performance of an app. App reviews work and feel more like postmortems, rather than a live feedback tool. As a result, developers will often find out what's wrong with their app too late, through users' negative feedback.

Postlaunch Challenges

One of the least talked about areas of developer activity is what happens after the app launches and the challenges this phase poses for developers. Developers have a number of issues to address, such as how to maintain applications after they have gone into production, pinpoint errors and bugs in your app, identify what the true user experience is like, and communicate directly with your customers, ensuring customer satisfaction.

Monitor and Maintain

Being able to monitor and maintain your app after it's in production is an enormous problem for developers. There are significant variables to address when you are an app developer: operating system, version of operating system (i.e., Android Gingerbread vs. Jelly Bean), device (i.e., iPhone vs. Nexus Galaxy, vs. Windows Phone), app version (which app version has the person downloaded), operator (AT&T, Verizon Wireless, Sprint, etc.), Wi-Fi or network, and screen orientation (landscape, portrait, etc.). All of these variables affect how your app performs for your customers, and once you combine these variables, the app will perform in much different ways. Therefore, it is important to have a tool to track and monitor these types of behaviors and issues so you have a real-time view of your app's health.

The challenge of tracking bugs and errors has prompted the emergence of many companies to support developers. Services like Crittercism (*www .crittercism.com*) track app errors by monitoring crashes and reporting the type of error, platform, device, and environmental variables like location, time, and transaction flow. As such, they can provide useful insights to developers, helping them fix errors before they drive users away.

Crittercism monitored crash rates from January to April, 2012 and found that the average iOS or Android app crashes 2 percent of the time. The company also found that a large number of users abandon an app upon its first crash, a fact that highlights the significance and utility of error-tracking services for developers, especially for brands and businesses.

Updating apps is another thorny issue reported by 25 percent of app developers, irrespective of platform. Interestingly, the difference in the update process between iOS and Android has no impact on developers' attitudes, as both iOS and Android have their own update challenges. On iOS, the process requires full certification and approval by Apple, plus explicit opt-in by the user. On Android, the update process can be automatic and near instantaneous.

This, however, requires that users opt in for automatic updates for specific applications. In effect, these challenges with the update process on both iOS and Android increase the average application "age" and escalate both code maintenance and customer support costs for developers.

Map Out Your Maintenance

Assume the worst and you will also be prepared for it. The same goes for your app. Don't think that the work is over when your app is available for download. You obviously want to minimize these feelings of apprehension and have the peace of mind that comes with knowing your app is supported. After all, you need the reassurance that bugs will be dealt with as they arise, and they will arise. It's not a case of *if* the app needs updating; it's a case of *when* the app will need updating.

The Maintenance Contract

This is where a maintenance contract can play an important role in your app lifecycle. Some app developers may offer a standard maintenance package. Others prefer to draw up a maintenance agreement for each app they develop and negotiate the terms on a case-by-case basis to allow for factors, such as the size and complexity of the app, that will determine the service level they can offer and the total cost. The service level defines the amount of work that the app developer will commit to providing, if necessary, for a specified period of time.

You should discuss the maintenance agreement early on in the project so that you can get an idea of the potential costs that you will incur during the lifecycle of your app. However, it's a good idea to leave the nitty-gritty negotiation of the maintenance agreement until you are nearing the end of the project.

Why wait? There are a several reasons why later is better in this case:

1. Both you and the app developer will have a clearer idea of the complexity of the app, and the resulting maintenance cost will be far more accurate.
2. You will have built up a good business relationship with the app developer by the end of the project and they will be more open to negotiation.
3. If your relationship has soured during the project, which can happen, you may decide that you would rather have the maintenance of your app performed by someone else. Keep that option open by leaving the topic until last.

ALERT

Negotiate a plan with the app developer that suits you both. Keep in mind an annual maintenance contract for an app should cost somewhere in the region of 10–25 percent of the initial development cost of the app, depending on the complexity and the service level offered.

This maintenance agreement service level comes as a surprise to many people, so it is useful to be aware of this before setting out on your app development project. Knowing all the costs—as well as the potential costs around your app—will arm you to conduct negotiations that will play in your favor.

What can you get for your money? It depends. For example, an app developer may provide a service level of two complete days per month for a fee of $1,000. Be aware that this fee is not dependent on the two days being used, and usually the days cannot be broken down into hours or minutes.

So, if you request a fix that takes just two hours, that will still be equal to one day of your allowance used. Usually, unused days are not carried forward to the following month. And if you require more work completed outside your allowance, that work will be subject to the normal fee structure of the app developer.

Crash Reporting Defined

A crash is any time a consumer is using your app and it immediately shuts down. When this happens, consumers will tend to leave a bad review of your app, and only 20 percent are likely to open up the app again. This is a real problem for you as an app developer. People are reporting that your app is crashing, but you don't know why, what device they are on, what OS version they are on, what app version they are on, and so on. You go to Apple and Google's available tools, but they won't show you the intricate details of why your app is crashing. What should you do?

Get the Report

A new category of software called "crash reporting" enables you to view the live health and availability of your mobile app and proactively manage app performance after it is in production in the app stores. These tools help you get a real-time grasp on what your consumers are doing in your app, and in addition, if there are any performance issues, they will provide you with diagnostics that include rich metadata so you know right away if the app has crashed or not and why. This way you can make data-driven decisions, focus engineering resources, and end mobile fragmentation headaches.

There are many tools on the market that can help you manage your crash reports. For example, with Crittercism, you can create a custom dashboard of relevant app metrics across app loads, errors, crashes, issues, and bugs, as well as high-level trends across operating systems and devices that track the performance of each app release over time. With these tools, you can set up real-time alerts to receive immediate app performance notifications so you can proactively manage any issues within your app, customize alerts to manage app performance, and receive real-time notifications to improve app health that will ultimately help you improve ratings, reviews, and revenue.

Some sophisticated app developers will add a "try-catch" scenario in their code. What this does is, when there is an error that happens when someone is using the app, it will move to another line of code before the app crashes. This is called an "error." These services also allow you to capture and track disruptive errors that interrupt the flow within the app even if the error doesn't result in a crash. This will allow you to trap these problems within your code and ensure that the app doesn't crash and allow you to debug your app quicker, while maintaining a good user experience.

See Customers in Real Time

Crash reporting also allows you to see what your customer is doing while they are interacting with your app, as well as identify what they were doing moments before a crash occurred, so you can see what problems you are having in your code and fix them. These tools are typically called bread crumbs: You can add additional debugging statements to your code, and see exactly what the user was doing right before a crash so you can re-create the issue. This will allow you to set up and track events to view app usage, trace user behavior to errors, and view all of the events happening within your code so you can respond quickly to issues, ultimately reducing investigation time so you can pinpoint and focus on the problems that really matter.

This unique approach allows you to focus on the most important issues, pinpoint why your app is crashing, analyze user behavior, and track custom attributes in order to optimize app performance.

You can also communicate directly with the people that are using your app. You can segment your list of customers that you send communications to on different parameters. That way consumers can provide immediate feedback when they face problems with an app, so they can immediately address the problem, improving user retention, and app store ratings and reviews.

Why Crash Reporting Is Important

Protecting apps from crashes is not only important for app developers, who by definition live and die by the ability of their apps to work smoothly. Increasingly, many types of companies rely on apps to run their businesses, even if there isn't a large consumer user base using the apps. Banks, for example, have mobile apps that enable check deposits; salespeople use apps in the field; and so on.

Competitive Advantage

Crash reporting and application performance monitoring is extremely important because it allows you to create a competitive advantage. When you have a bird's-eye view of what is wrong with your app, you are able to address approximately 75 percent of your issues by pinpointing, identifying, and fixing your top ten errors. This in turn will help you eliminate one-star ratings so you rise on the leader board in the app stores. Also, if you can proactively reach your customers and notify them that they need to upgrade their app, you can catch stability issues before consumers leave that one-star review. Ultimately you are able to triage the most critical issues with a data-driven approach.

Save Money

Once you know what issues are happening in your app, you can identify stability issues that are leading to end-user drop-off so you can stop losing money. Once you have visibility into the issues you are having with your app, you will be able to address them quicker, therefore optimizing engineering resources and reducing development costs and headaches. You will now be able to establish clear, transparent metrics on code quality, improving your developers' QA process post–app release.

ESSENTIAL

In addition to having a real-time view into what's happening in your app, you are able to make data-driven decisions by having deep insights into your app metrics. You can see the breakdown of operating systems, usage, and devices, and funnel development resources and investment into what is working for your particular app.

Reasons Why Apps Crash

It turns out there are many possible reasons why specific apps crash. As you work to maintain your app, you should try and be aware of as many of these reasons as possible, especially if your operating system is prone to more issues than others. Crashing can vary, particularly depending on whether you are using an Apple iOS device such as an iPhone or iPad, or an Android device.

Too Many Platforms

One of the reasons for app crashes is the proliferation of mobile operating systems on iOS and Android. As Apple and Google have released more new operating systems, each with multiple updates, app developers face more operating systems to test their apps on. Crittercism found that between December 1 and 15 of 2011, there were at least twenty-three different iOS operating systems on which apps had crashed and thirty-three Android operating systems.

The largest proportion of app crashes from both iOS and Android platforms was on iOS 5.01, with 28.64 percent of overall crashes (in a normalized data set). This makes sense, as iOS 5 was still relatively new at that time and many apps still needed to work out the kinks with the new version.

There are also older iOS versions that have a significant number of app crashes. For example, iOS 4.2.10 had 12.64 percent of app crashes, iOS 4.3.3 had 10.66 percent, and iOS 4.1 had 8.24 percent.

What Causes a Crash?

So why do apps on these operating systems crash so much, and do iOS apps crash more than Android apps? On the first question of why apps crash, the reasons are many. A crash can happen because of hardware issues, such as the use of location or GPS services or cameras; it could be due to the Internet connection—that is, how a phone connects to 3G or Wi-Fi—or that the device is not connected to the Internet at a certain moment, or that something happens during the switch between 3G and Wi-Fi. There could also be issues with language support on certain devices. There can also be memory problems if an app uses too much memory.

Problems can also occur with third-party services that developers use in their apps, from analytics to advertising systems. For example, there were reports that Apple's iAds system gave some developers problems if they did not adhere to certain standards.

People don't update their apps very often, just as they don't update their operating system. (Android, unlike iOS, allows users to auto-update their apps, which can eliminate some of the problems.) So developers often test all previous versions of their apps with each version of the different operating systems. Particularly with a new OS platform, developers have to test their app to make sure it still works.

Android vs. iOS

Crittercism found that Apple iOS app crashes accounted for more of the app crashes than did Android-based phones. In addition, Crittercism analyzed a total of more than 214 million app launches from November and December 2011, from apps that use its service, and found that there were about three times more app launches for iOS than Android, about 162 million to 52 million. (Note that the analysis examined app crashes as a percentage of each app launch, so this data takes out the issue of there being more iOS than Android apps.)

In the top quartile of apps, Android apps crashed 0.15 percent of the time they launched, while top quartile iOS apps crashed 0.51 percent of the time. In the second quartile of apps, Android apps crashed 0.73 percent of the time and iOS apps crashed on 1.47 percent of their launches. In the third quartile of apps, Android apps crashed 2.97 percent of the time, while iOS apps crashed 3.66 percent of the time.

On a basic level, iOS apps crashed more than Android apps during this time period. But this doesn't necessarily mean that overall iOS apps crash more than Android apps. That's because Apple had recently released a new version, iOS5, in October. Android's new Ice Cream Sandwich operating system (Android 4.0), meanwhile, had not been widely released on phones yet at the time of this study. Still, the numbers can't be ignored: The data shows that apps on iOS did crash substantially more than Android apps.

When you look at the top apps in Crittercism's data, it doesn't really make a difference that Android apps have a lower crash rate than iOS apps because they are both well below 1 percent. However, there was a bigger difference between iOS and Android app crashes in the top quartile of apps than in the third quartile. In other words, the best Android apps crashed about one-third as many times as the best iOS apps, while in the second-best quartile, Android apps crashed about half as much as comparable iOS apps. In addition, the very top Android apps are achieving a crash rate that, at least in this time period, the best iOS apps can't match. Why is not entirely clear.

However, Android, it should be noted, allows developers to push updates faster than Apple. With Android, developers can just send an update to its code, which can show up almost in real time. But for iOS it can take days or a week for an update to show up. That means there can be more app crashes while those updates are waiting to happen. Whereas with Android, presumably, if developers know there's a bug, they can immediately fix it.

FACT

The performance of apps is not only different on various operating systems, but also on different devices. About 74.41 percent of the iOS crashes Crittercism tracked were on the iPhone, 14.81 percent were on the iPod Touch, and 10.72 percent were on the iPad.

Platform Problems

The primary operating systems for developers are Android and iOS. As Apple and Google launch new versions of their operating systems, it is difficult for developers to keep up with all of the improvements offered in each of these new versions. It is critical for developers to have a real-time view into what is happening with their apps, pinpoint when and where their apps are crashing, discover and fix the specific causes of problems, analyze user behavior, and track custom attributes in order to optimize app performance.

Android and iOS

With the launch of iOS6, Apple has released a number of new and exciting features to improve the consumer experience, such as improved mapping features, face recognition, single sign-on for Facebook integration, boarding passes, and more. These new features will also impact mobile app performance.

In addition, Apple has released the iPhone 5 with many new device improvements. While these new features can be tested, it's difficult for developers to understand how the app will perform once live in production. With crash reporting tools, developers can prioritize and fix top issues immediately, acting as an insurance policy against poor app reviews.

With the launch of Android 4.1 Jelly Bean, Google launched Google Now. The enhanced notifications, improved calendar and browser, and the overall speed of the user interface are just a few things that make Android

4.1 Jelly Bean operating system one that owners are desperately craving. Problem is, aside from Nexus devices and a few others, Android 4.1 Jelly Bean remains scarce in the United States. In fact, according to Google, a mere 2.7 percent of devices currently run Android 4.1.

For Android, due to fragmentation of the operating system and devices and the impact this has on app performance, it is critical that developers have visibility in what is happening in the apps and what issues they are experiencing so they can quickly fix problems in the code.

FACT

Android fragmentation continues to be a major headache for developers. To put this into perspective Open Signal Maps (*www.opensignal.com*), a company that creates crowd-sourced wireless coverage maps, showed the distribution of devices using their Android app. Drawing from data from 681,900 users, it found that there are 4,000 unique Android devices out there running its app and only a handful were running the most recent version. The point here is that developing for Android is no longer as straightforward as it once was. With so many devices and an ever-expanding catalog of operating systems to cater to, it's hardly surprising that Android support is shaky at best.

HTML5 and Hybrid Apps

Mobile app developers are embracing HTML5 in their app development processes. According to the recent survey by mobile app tools vendor Kendo UI, 94 percent of developers are either using HTML5, or planned to start using it in 2012, leaving only a minuscule 6 percent who have no plans to develop with HTML5.

HTML5 is an updated version of the old-school hypertext markup language that is found on much of the web today. It enables developers to build on their existing knowledge of web technologies such as HTML, Javascript, and cascading style sheets to create mobile apps through cross-platform frameworks such as Adobe's PhoneGap, rather than having to learn Objective-C to write full-native iPhone/iPad apps, or Java to write Android apps. Probably even more importantly, by using cross-platform technologies like PhoneGap, HTML5 enables developers to write their apps once and deploy on all major mobile platforms.

Developers' rationale for using and preferring HTML5 is no shock to anyone who's ever developed native apps for multiple mobile platforms. Sixty-two percent said that HTML5's ability to enable cross-platform support was an important factor in choosing the technology, with another third saying that the availability of tools and code libraries make it appealing.

ALERT

HTML5 is the ubiquitous platform for the web. Whether you're a mobile web developer, an enterprise with specific business needs, or a serious game developer looking to explore the web as a new platform, HTML5 has something for you!

Keep Your App Healthy

Whether you focus your app development on Android, iOS, HTML5, Windows Phone, BlackBerry, or any other platform, it is important to have open visibility into your app's performance issues and errors so that you can provide the best possible user experience to your customers. There are many options to choose from, but the most important factor in your selection of a platform is that it should meet your needs and demands and provide you the tools necessary to manage and maintain your app.

Sections of this chapter were contributed by Jez Harper, Tús Nua Designs.

CHAPTER 18

Success Breeds Success

Have you ever wondered what aspect of the mobile app industry will have the biggest impact on your life and business? Will it be the convenience that apps bring or the services they enable? Perhaps it will be the enhanced power the consumer has over the retailer or the complete transparency of pricing. From a developer standpoint, maybe it will be the expanded influence of the "open API" beyond web-based apps. Whatever this nascent industry transforms into, the one thing that has clearly been impacted by the opportunity mobile has brought is a new breed of entrepreneurship.

The Rise of the *Appreneur*

From the lush rice fields of Indonesia to the barren desert of western India and right across the entire planet, mobile technology has unleashed a torrent of appreneurs, democratizing business on a global scale. There really hasn't been a time in history that has given voice to more people to build more companies, grow economies, and do more good than right now. Mobile technology isn't disrupting business, it is reinventing it. Just as the automobile went from a horse and buggy replacement to the instigator of the invention of fast food and the highway system, business is ripe for a redo, all thanks to mobile.

ESSENTIAL

Appreneur brings together the words *app* and *entrepreneur* for a reason. Appreneurs are not hobbyist developers, they have built a business. Good appreneurs are smart about how they structure and build their business, how they choose and manage their development team, and ultimately how they market and monetize their apps.

Appreneurship 101

So what does it take to become an appreneur? It is one thing to learn the ins and outs of the process of building, deploying, and marketing your idea, it is a completely different thing to execute the idea properly. This industry, like so many revolutions, has been built by the trials and errors of the crazy ones who set out on this journey before you. They are the ones whose lessons and experience will guide you as you build your own app business. The following are some valuable, real-life examples of the trailblazers that are reinventing business through mobile. The companies and individuals collected here have lived this industry's ups and downs and have taken different approaches to arrive at one goal: achieving success to ensure sustained market presence.

Project NOAH: A Case Study

Have you ever looked at an insect or a plant and wondered what it was? Perhaps it is a new species; maybe it is poisonous. Not long ago you would have to look it up online or in a book, but there is a company that is transforming the way all of the world's organisms are classified and identified, and they are doing it through the use of mobile technology and a little human ingenuity.

If Darwin Had Invented Foursquare

The company is Project NOAH, which stands for Network of Organisms and Habitats, and it was launched as an experiment at New York University's Interactive Telecommunications Program in early 2010 to be a "virtual butterfly net." Since then, they have classified over 300,000 organisms from over 100,000 users around the world and are now supported by the venerable National Geographic Society.

Yasser Ansari, founder and "Chief Leaf" at Project NOAH, realized very quickly that the best way to collect and classify the planet's organisms was to make it easy, make it convenient, make it fun, and use something that most of the population carries with them, the mobile phone.

If just a small portion of the population that carried a device (over 6 billion devices worldwide) snapped a photo of a living organism and posted it online for classification by hobbyists and experts alike, using this network effect, they could create the most comprehensive database of things that live on earth. Not only that, they could then use the data, collected continuously over the years to get a better handle on wildlife patterns, species reach, migration, catastrophe, and disease. This data could then be used to take Mother Nature's pulse.

Overnight Sensation

The biggest challenge Project NOAH creators faced was getting the app into users' hands. This was compounded by the fact that they released a version of their app to the Apple AppStore with the wrong minimum technology specifications, so the app didn't load for a number of their early customers.

Because of the submission and approval process that Apple adheres to, days passed before the app was updated, and many thousands of users downloaded software that didn't work. Contrary to what you'd expect, the die-hard nature nerds out there banded together. Instead of roasting the software company, they spread the word about the app on their own, using social media and word of mouth.

Once the Project NOAH team's latest release was approved, there were a large number of eager people who immediately downloaded the app. The result was an overnight sensation that would likely not have happened had the developers released the app properly the first time. It's one time that it actually paid not to be perfect. The increased demand also helped make the app a "new and notable" app in the AppStore, an accolade that created even more awareness. While this was not a conscious strategy, it does show the power of creating anticipation for your app launch.

Media awareness was also a key challenge for the Project NOAH team. How could they rise above the noise to make sure traditional larger media would notice them? The "perfect storm" of pervasive computing and accessibility also created a misconception that much of today's technology keeps people indoors, glued to a television screen or computer monitor. Project NOAH was able to use the fact that, in order to use their app, kids needed to be outside. After the media picked up on this angle and ran with it, the app was featured in some high-profile publications and outlets.

Mobile App Lessons Learned

The Project NOAH founders resisted the temptation to think big too soon. Many times the conversations within the team moved to building other tools that leveraged the Project NOAH platform. This kind of distraction can hurt a company fighting to survive, and the founders understood this very early on. They did the wise thing and put these discussions on the back burner until they saw this first phase through.

The singular focus paid off, and today the founders are receiving offers to have their entire platform white-labeled to provide the basis for other initiatives. The lesson: Don't try to be everything to everyone by doing too much. There were many opportunities for Project NOAH to expand beyond the core focus of cataloging nature, but they turned them away in order to focus on doing this first initiative well.

They also learned some other important lessons:

- Find a niche that is underserved in the mobile space. Project NOAH leverages the intersection of three emerging trends: mass adoption of mobile devices; the anti-screen movement thought to be brought on by technology; and the networked, always connected and accessible world that is emerging as a result of mobile communication.
- Create a succinct message that is repeatable and encapsulates your reason for being. Project NOAH is taking Mother Nature's pulse and allowing people to connect with nature using a digital "butterfly net." The founders have delivered a specific message to a well-targeted audience.
- Partner with someone who has an audience. Whether it was done on purpose or not, the fact that Project Noah teamed up with National Geographic added instant credibility to the app and—more importantly—exposed the app from the start to the nature-loving National Geographic audience. Reach and speed are important for an app, so you are well-advised to consider forging a strategic partnership to promote your app to your audience.
- Don't be intimidated by the Big Guys. Despite being underfunded and lacking the clout of the incumbents, Project NOAH managed to create an app that attracted more users, more engagement, and more media attention than any of the rival companies, some of which have been around for over 100 years.

The founder of Project NOAH understood the need to get all of the aspects of the app in line and not just the technology: weaving a story and selling it, finding an awaiting audience without a good enough mobile offering, leveraging the networked world and the pervasiveness of mobile phones. At a more human level, they correctly jumped on the do-it-yourself/get-involved trend that is sweeping across a number of industry sectors.

ALERT

Building an app is not as simple as lines of code. To be a successful appreneur, you need purpose, focus, and a clear idea of the value you provide.

LevelUp: A Case Study

Two years ago business partners Seth Priebatsch and Michael Hagan started a location-based game called SCVNGR. The concept was simple but grandiose: to add a game layer to the entire planet. They were one of the first companies out of the gate and managed to raise funds from Google Ventures as they set out upon their mission.

Two years later, SCVNGR is gamifying the world. They also started to listen to the feedback their customers were providing. Their clients, retailers in particular, wanted the apps to encourage deeper loyalty with their customers. The aim: use the app to keep consumers coming back to the stores. Based on feedback, the business partners launched LevelUp, a company focused on mobile payments and mobile loyalty.

LevelUp is trying to solve the "interchange" problem, that is, the cost borne by merchants interested in offering their customers the ability to pay for goods with a credit card. What makes this company, and their app, stand out from the mass of other mobile payment apps is their sharp focus on customer retention and loyalty.

New Thinking, New Markets

The way LevelUp works is simple for the merchant and the consumer. The consumer downloads the LevelUp mobile app and links their credit or debit card to their phone. They can then use the app to pay their bill at participating retailers. The differentiating factor for the merchant is there is no interchange—the fee that is charged by the credit card companies to the merchant. Instead of charging that fee, LevelUp converts it into a discount on the consumer's next purchase.

The founders overcame some considerable hurdles in order to shift gears and reorient their focus to mobile payments and loyalty from the gaming company they started with. After all, location-based gaming was core to its business DNA. It takes a disciplined team to re-evaluate a successful business platform that was gaining considerable traction and be critical enough to realize it could never scale the way they had anticipated.

In going through this exercise, they discovered the pain that their customers were feeling and set up a new company to solve it. Even with this firsthand information, they launched with a product that, in hindsight, wasn't completely attacking the pain. So they went back to the drawing board and through a lot of trial and error and customer involvement, they landed on the company's current offering.

Help Your Customer Succeed

If you are in business, you've probably heard the adage "Be the aspirin." This means be sure that your app is a must-have, not a nice-to-have. The founders realized their first endeavor, SCVNGR, was cute, but not critical in their customer's lives.

They used this concept of cute vs. critical to govern their decisions about LevelUp from that point forward. They mined their customers, asking what their primary pain was (they did not assume what it was; they asked), and homed in on the factors that could change their business, while at the same time becoming an essential service that their customers couldn't live without.

Through this process, they realized that by solving the high interchange costs for merchants (pain #1) they could also boost consumer loyalty (pain #2). When LevelUp becomes engrained in the business of shops and retailers (and the habits of their customers), then it's tough for competitors to rise up and remove LevelUp from the equation.

Change When Needed

The process of transitioning a business comprises a number of decisions that need to be made along the way that help paint a clearer picture. There are a finite number of decisions that can be made that will alter the direction of the business. The key is to *make* them. Every day that passes, every decision that doesn't get made gets made for you by the very fact that no decision was made.

ESSENTIAL

Ultimately, having decisions made instead of making them means the business owner is not in control, is not running the ship, and will likely be out of business sooner rather than later.

The founders of LevelUp started making decisions about the future direction while they were riding high with a successful and "loud" product in SCVNGR. They were constantly surveying the landscape, talking to their customers, understanding the trends, and consistently questioning what they were doing—even when things were flying!

Listen to Customers, but Ask the Right Questions

Questions are easy to screw up, especially when asked in the context of a service or product currently in use. Think about the questions you ask your customers every single day. Are they asked with your current offering in mind? Are they evolutionary, or revolutionary? Are they questions about improvements to the current offering, or are they questions about what you should be offering?

When the founders of LevelUp approached their existing customers and new ones, they asked the questions that led them to understand the real pain they were feeling. Assumptions can be company killers. Find your customer's pain(s) by asking the right question.

The reason questions are so difficult is that the answers are often clouded by the current situation. A business owner answers questions differently based on factors within the context of their business and the service they are providing. It's your job to ask and ask and ask until you break them out of the current process and get them thinking about what they should be doing.

Simplify Your Offering So Your Mom "Gets It"

Your mother should be your litmus test for any business. Can you distill your message to one sentence that your mother or father would understand and be able to explain to their friends? You need to do this, because this forces you to make it very clear what your app does and why anyone should want one in the first place. LevelUp stuck to the basics. SCVNGR is a business that adds a game layer to the world, and LevelUp enables credit card transactions while making customers more loyal and thus spend more.

Your business might be something you understand, but is the way you describe it simple enough for you to play the telephone game with a room full of retirees?

Never Stop

Once you think you've got it, keep going and don't stop. Think of yourself as a basset hound. The beauty of basset hounds is their perseverance. Once they stumble onto a scent, they won't deviate from the trail until they find where it is coming from. No distractions. No detours. Just a single straight line from the scent to the end of the trail. They have been bred for certainty.

Most people lack that singular focus. They tend to overthink, overanalyze, and ultimately fail to follow the appreneur instinct. Don't give in. If you, like LevelUp, have talked to customers about their pain, have established that you are critical (not cute) to the success of their business, and can successfully communicate your business proposition to your parents, then you have what you need. Don't waste time. Develop or market that app!

The transition the founders of LevelUp went through from the launch of their first product, SCVNGR, to the launch of their second product, LevelUp, required them to change their mindset completely. Through trial and error, customer feedback, and their understanding of the industry and where their customers needed immediate help, they were able to offer a mobile product that was easy to sell because it solved real-life problems.

The key to success for them was allowing their customers to make more money. A simple focus like this guides decisions. If you help your customers make money by using your app, they will likely never leave your side. What's more, they will always help make your product better and will advocate on your behalf.

Gigwalk: A Case Study

Mobile has triggered a shift in the way people find and deliver work. Contextualized work, work that is location and time dependent, has always been a challenge to staff up and execute efficiently. Until now. An app using a location can now open up working opportunities as never before, and one company has helped countless unemployed or underemployed workers find a new source of income via their smartphones.

Contextual Work

Gigwalk is a mobile work marketplace designed to help match available work with people who can do it, all based on their location. Employers looking for part-time or temporary workers post a "gig" in the system, tag a location for that gig, what they are willing to pay, and the timeframe for getting it done. Users of the Gigwalk mobile app can then find work based on their location and available time.

ESSENTIAL

Gigwalk businesses have posted over 190,000 gigs across the United States, helping over 140,000 people find work and increase their income, all through their mobile device. Coordinating and managing something on this scale presents a logistical and financial challenge.

How does a company mobilize and manage a temporary workforce and keep costs down while ensuring the quality of the work being performed? Think about the process that would have been in place prior to a company like Gigwalk. The employer would have had to post a job, accept resumes, vet the resumes, hire, explain, train, and execute the work. Then, if this was a national or international initiative, the employer would have to do it all over again for each individual location. Then they have to manage the whole thing and hope that the people they've hired do their jobs properly. A logistical challenge to say the least.

Gigwalk was started to ease all of these challenges. They put it together one small step at a time. First, they opted for an extensive beta period of close to one year to test their app with users. It might seem like a long time to spend in beta, but the information they were able to gather over this period helped them figure out how to rate the workers, determine the most effective jobs to post, and lay the groundwork for a database of employers for their app launch.

Once they launched, they did it in a market they understood, and they stayed hyper-local by focusing on neighborhoods, not nations. Gigwalk launched in a neighborhood in Los Angeles in a niche they understood in order to continue to test how this application worked in the wild. It was important for them to build the app in a way that could scale, extending their reach. The app also had to manage itself, so it was critical to debut their app in a location where they could also visit the employers for critical feedback.

A lot of what they did during the beta period was test theories and change the things that didn't work. This was key to their successful launch. The approach was also flexible. The company guessed that if they didn't dictate to customers how they should use the app, then these same paying customers would provide invaluable input on how they really wanted to use the app.

Mobile Lessons Learned

The outcome of this prelaunch effort is what created a successful launch and was critical to ironing out the potential issues that would arise on a national rollout. What other mobile app lessons did these successful appreneurs learn?

Let your customers' imagination run wild. The founders at Gigwalk had an idea about what their service offered, but they also had the sense to let their customers think about the possibilities beyond the original vision. When a customer starts thinking about a product in that light, they are connected to the idea and will find a reason to use it. Test and then test again. Gigwalk spent an extended time in beta to make sure they ironed out all the details and processes they needed to in order to launch across the country. Never rush the launch to get to market, as this may sacrifice your ability to scale quickly.

Define your customer base. The industry that Gigwalk is a part of is very competitive, especially on the consumer side. Despite the year-long beta effort they did before launching their product, Gigwalk has now shifted their focus to the enterprise market to help employers there. Thus, Gigwalk is moving into new and potentially lucrative territory. The enterprise market is underserved, and Gigwalk can fill that gap quickly. After all, the company spent two years perfecting their app. It's smooth sailing.

Simplify your offering. If your aim is to change the status quo and established habits, then don't be complex. Start simple and move your customers along gradually on the learning curve. Offering too much change or an app that does too much will be hard for customers to adopt.

Gigwalk had a vision of changing the way people work based on the few things they take for granted: location and the scarce resource of time. Mobile has opened up the opportunity to finally build a new approach. It's also opened up a discussion around the future of work.

From Tweets to Living Photos

There's no lack of opportunity in the mobile app space. The trick is to learn how to separate the surefire ideas and approaches from all the noise. It's all about following a strategy that allows you to identify a true sweet spot that allows you to build an app business and prosper. This requires patience—the kind that tests the balance between passion and pragmatism. It's also what separates the leaders from the also-rans.

Shifting Focus

You've surely seen this scenario play out: An entrepreneur forges ahead with a great idea, only the rewards don't materialize because being a success was dependent on them being a fast mover with no followers. They counted on being the only one in their space. There's a big difference between observing that other companies are in your space, and recognizing it may be time to move on.

This is exactly the moment of truth that impacted the app business belonging to Bretton MacLean and Mark Pavlidis. They launched Tweetagora, a Twitter app for power users, a few years back, only to see their business evaporate when Twitter moved into their space through the acquisition of Tweetie. This marked the beginning of the end of the developer support provided by Twitter.

ESSENTIAL

The "open API economy" is great to get things off the ground in a very inexpensive way. There hasn't really been a time in history where you could build a product so quickly and for so little money.

As Mark and Bretton found out, there will come a day when you need to realize that "cheap" has a downside, and building an entire business on someone else's technology means that a single change or acquisition on their end could destroy the value you provide to your customers.

Think ahead and answer this question: What happens if your provider changes their model, and what is the likely impact on your app business? What did Bretton and Mark do? They struck out on a new journey to look for a newer, larger opportunity. They found it—and got in just before the growth curve took a swing to the sky.

Shifting focus from Tweetagora to Flixel was a necessity, and it teaches a valuable lesson: If the core of your business belongs to someone else, it isn't your business. Why? Because if the core shifts, the business is lost. In the case of Tweetagora, Twitter changed their rules and the impact was devastating for Bretton and Mark's original business idea. But, as true mobile entrepreneurs do, they picked themselves up, dusted themselves off, and got down to work on their next app: Flixel.

The market at that time was populated by a variety of apps, including Instagram, that allowed people to take and share photos. At the other end of the spectrum, there was a rush of companies moving into mobile video sharing. There was no middle ground, and no room for differentiation, or so it seemed.

But this was because other companies were missing the opportunity in cinemagraphs (a mash-up of still photos with one part of that photo being animated in a smooth loop to bring the photo to life for one moment). Cinemagraphs are a perfect hybrid between the photo world of Instagram and the video world where app developers and companies are jockeying for position. The big difference is that: cinemagraphs are part of what consumers already know and appreciate: "living" photos.

Flixel is an iPhone app that lets people do precisely this—quickly and easily. Traditional photos are completely still; a living photo contains a portion of seamless and infinitely looping motion. Whether it is hair blowing in the wind, or a flickering fire in the background of an otherwise still image, the effect can be mesmerizing.

Today Flixel is more than an app; it's a premier platform for discovering, creating, and sharing living photos. Users download the app, record a short scene, "paint" the portion of the image that will animate, then share it on a variety of social networks including Twitter, Facebook, Tumblr, and Flixel's own community.

Mobile Lessons Learned

There are multiple lessons you can learn from this story. First, build a product, not a feature. When Twitter bought Tweetie and then bought Tweet-Deck, it destroyed thousands of businesses instantly. Don't put yourself in the situation where your product can be replaced that quickly and effortlessly by incumbents. Build a product, not a feature of someone else's product.

Avoid targeting super users as your first customer. Super users may be the worst type of customer to target! They are early adopters and jump ship when they get bored. Better to go after the customer that takes their time to make a decision, as they are more likely to stick with their choice.

New Rules, New Game

The bottom line: The app business is just that, a business. But this isn't entirely business as usual. Granted, the rules of Retail 101 still apply, but you also have to think mobile. What does it take to be an appreneur? It takes a desire to change something for the better and the ability to finish what you started—to execute. Regardless of the platform or the revenue model or the marketing tactics you choose, there is one thing that is for certain and that is business is changing at the hands of the appreneur. And who knows, perhaps what you build after being inspired by these pages will be the subject of its own book one day.

Sections of this chapter were contributed by Rob Woodbridge, Untether TV.

CHAPTER 19

The Future of Apps

The advance of mobile devices and the abundance of apps are coming together to transform the daily lives of millions of people. From apps that enable more "human" customer service, to apps that pave the way for more effective and personalized health care, to personal digital assistants that provide us advice every step of the journey, to apps that mobilize the enterprise, it's clear that the next wave of innovation will focus on utility, not novelty.

Looking Forward

As independent consultant Chetan Sharma observed in his milestone report on the App Economy called *Sizing Up the Global Mobile Apps Market*, "connectivity breeds apps." As he predicted, the industry is witnessing a significant uptick in the apps that extend beyond smartphones and tablets to include cars, digital cameras, personal navigation devices, picture frames, weight scales, and the list goes on.

This trend is confirmed by new research numbers from Appcelerator (*www.appcelerator.com*). When Appcelerator asked 5,000 app developers to look into their crystal ball and share their vision of a connected world (and the role they would play) in 2015, they found that developers predicted it is "likely to very likely" that they will be building mobile apps for more than smartphones and tablets. Televisions (83.5 percent "likely to very likely"), connected cars (74 percent), game consoles (71.2 percent), Google Glass (67.1 percent), and foldable screens (69.1 percent) all ranked high on the list of future form factors. Despite the entry of a variety of additional form factors, many developers still believe they will build the majority of their apps for smartphones and tablets in 2015.

Apps Power Machines

The opportunities for apps also extend to the M2M (machine-to-machine) market, where key verticals including telecom service providers, utility companies, hospitals, and logistics companies require apps to extend the reach and appeal of their services to mobile devices.

ESSENTIAL

The overall M2M market encompasses an extraordinarily broad range of areas where apps can add value, including automotive telematics, smart metering, and smart cities. In these smart cities sensors, networks, and intelligent systems help authorities manage water, gas, electricity, waste, transportation, and fuel use, providing urban dwellers an improved quality of life.

Automobiles

Automotive is once again where the action is. The first wave of applications were focused primarily on the vehicle, improving the in-car experience by enabling accurate navigation, the timely delivery of turn-by-turn directions and automating in-car entertainment.

The industry appetite for apps is no longer limited to enabling basic navigation and telematics services. The opportunity has grown, with car makers such as BMW creating entire app develop communities to encourage the proliferation of apps to make cars smarter. From apps that enable the car to "find" a free parking space, to apps that help road warriors manage their businesses on the move (and from their car), the race is on to equip cars with new capabilities and intelligence.

ESSENTIAL

BMW, Mercedes-Benz, Audi, Honda, and GM are among the automakers integrating apps into their vehicles. So far, Microsoft's SYNC for Ford vehicles is leading the industry in functional apps and in pushing the envelope.

And it doesn't stop there. Brands and marketers are also beginning to view the car as another screen where they can deliver advertising, marketing, and more. In this scenario, an app could deliver Groupon-style local deal offers to users as they drive by a business, and then sweeten the offer with turn-by-turn directions to get to where the bargain is.

Another area of opportunity is for apps that help package and personalize content for in-vehicle enjoyment. Imagine an app that delivers a tailored newscast or music mix that features only the stories, topics, or genres of most interest to the driver or passenger . . . *personally*. No more listening to the standard fare; the app will "decide" what matters most.

Or, imagine a situation where the driver is in the car on the way to an event and doesn't have the print invite. An app can function like another access point to everything stored at home or at work and serve up the electronic invite and any other pertinent information. These are just a few of the scenarios where automakers say apps will fill the gap.

Apps Improve Customer Service

Consumer research shows people look at their mobile phone screens between 40 and 140 times a day. The jury is out on the exact total, but there is no arguing against the pivotal role mobile plays in people's daily lives. Time-crunched consumers are looking to their apps as ways to remove friction from activities such as paying bills, booking flights, and making reservations. Against this backdrop, the pressure is on companies across *all* sectors to deliver customer service that is not only more flexible, but also more convenient and personal. Anything else will likely be dismissed as a waste of time.

Be Flexible

Being flexible means adding mobile apps to the mix of ways people connect with companies when they have issues or questions. Of course, there's no single best answer when it comes to consumer preferences. Some want to speak with a call center agent; others want to perform self-service tasks using a mobile app. And a significant and growing group of consumers want it both ways. They want to perform self-service tasks on an app and connect to an agent (without having to hold) when an issue can't be resolved using other channels.

To validate consumer preferences across all demographics Vocalabs (*www.vocalabs.com*) conducted a survey of 900 smartphone owners on behalf of Nuance Communications (*www.nuance.com*). Mobile apps were rated as an important part of the self-service mix that defines excellent customer service. Almost half (45 percent) of consumers surveyed said they like to use customer service apps because they are convenient. Another 40 percent like the always-on nature of mobile apps, and the fact that they are "always available."

When it comes to customer service inquiries, 75 percent said they find self-service to be more convenient. Furthermore, if the self-service experience is positive, the majority of participants said it triggers a perspective that the company they're doing business with is customer focused and innovative.

Pinpointing Customer Service App Shortcomings

Research from Nuance and Vocalabs shows that not all verticals offering customer service apps are doing a good job of it. Significantly, banks and mobile operators lead with the largest number of app downloads. However, these same verticals stand out as the companies that experience the most serious drop-off in usage. While 60 percent of respondents have downloaded customer service apps from mobile carriers, only 25 percent are actually using them. That means over one-third (35 percent) of respondents are *not* using the apps. Banking apps show a similar disconnect.

ALERT

The serious shortcomings in banking apps is the topic of a new report by My Private Banking Research, headquartered in Switzerland. Based on a survey of 350 users and an audit of some 300 mobile apps, the report confirms that banks are failing to meet customer service requirements with their mobile apps.

But it's not just about apps that equip people to conduct banking transactions or find the nearby branch or ATM. The report shows there is also a demand for app features that allow people to move directly and seamlessly from a mobile app to a real-time conversation with a bank advisor.

What features and functionality could convince consumers to use their customer service apps more? Nuance consumer research highlights areas where improvements would pay dividends. Over one-third (35 percent) of respondents would appreciate a seamless and effortless way to shift from a self-service task on their app to connect with a call center agent.

It's all about the freedom to choose. If consumers can't accomplish what they want within an app, they have to disengage and try another channel.

Voice Raises the Bar

Mobile technology empowers people to manage their lives and get advice and information on the move. But it is voice that has the potential to remove all the obstacles, allowing people to get more done, faster. Rather than scrolling, clicking, and using traditional search to find what they need when they need it, people can ask their mobile personal smart assistants to lend a helping hand.

Siri, the smart assistant service available on Apple iPhone 4S (using technology licensed from Nuance Communications) is a perfect and powerful example. It combines speech recognition technology with natural language understanding (the ability to understand the simple spoken word, rather than stilted, unnatural voice commands) and tight integration with key device functions (such as contacts and calendar) to usher in a new era of communication.

ESSENTIAL

Siri is the game changer that has raised consumers' awareness of the power and simplicity of voice for getting instant answers and advice.

Siri helps people make calls, send text messages or e-mail, schedule meetings and reminders, make notes, search the Internet, find local businesses, and get directions. Consumers can also get answers, find facts, and even perform complex calculations. All people have to do is grant Siri access to the personal data and details that make up their daily digital lives.

Services like Siri and Dragon Go!, the mobile app from Nuance that leverages both speech recognition and natural language understanding, do more than create a customer requirement for increased convenience. They also set a high standard for how people expect to interact with their apps.

Put another way, the advance of Siri and services like it means companies will need to harness natural language understanding and voice recognition to deliver their customers information, entertainment, advice, customer service via websites, and apps.

The explosion of mobile voice services and the advance of smart mobile voice assistants are already having a significant influence on how consumers interact with companies. Moving forward, speech recognition and natural language understanding are poised to be an even more powerful combination, adding another component to the toolbox of capabilities app developers will need to meet and exceed consumer expectations for apps that help them manage their daily lives.

Artificial Intelligence Gets Simpler

People are relying more on their mobile apps to get things done, and apps are beginning to harness artificial intelligence (AI) to rise to the challenge of enabling more complicated services.

AI is all about making technology behave like humans. The term *AI* was coined in 1956 by John McCarthy at the Massachusetts Institute of Technology and refers to a variety of approaches to making machines intelligent, ranging from game playing (programming computers to play games such as chess and checkers), to expert systems (programming computers to make decisions in real-life situations such as diagnose diseases based on symptoms), to natural language (programming computers to understand human language).

ESSENTIAL

When people think of AI, they often visualize film creations such as the HAL 9000 from *2001: A Space Odyssey*, and Skynet from the *Terminator* movies. But practical AI applications have nothing to do with scary movies about how computers can ruin our lives.

AI is about infusing technology with human intelligence; and it's about keeping that knowledge fresh through deep and ongoing interactions between man and machine. In mobile that interaction involves users, their devices, and their apps.

This close and constant interaction between people and their devices allows the device to learn about its user to the point that it actually knows what its owner wants, even without the owner having to ask. How is this possible? The device (and the mobile app running on it) is able to access important information and data by drawing on functionality and features baked into the devices, including location information, an accelerometer, and the camera. The device (and the app running on it) is able to understand user context and deliver what the user needs most.

So what characterizes an intelligent app?
It can be many things but, in essence, it is an app that does a task for a user like personalizing the news they access (because the app "knows" what content the user wants based on past experiences and has therefore aligned the content it delivers with the user's preferences, for example), or optimizing the user experience based on contextual information such as location and time of day.

The potential is huge, as there is a need for an intelligent app in most areas that involve search and choice; that is, for every service. To name a few examples, you can use AI to build: a micro search engine, a recommendation app, a shopping guide, interactive customer care, a dating app, a virtual assistant, a recruitment app, a discovery solution, a mobile city guide, a banking assistant, consultants of selling and configuration, product comparison and matching, and much more.

AI Tackles Information Overload

The requirement for more intelligent services will likely accelerate as people become more sophisticated in how they use mobile devices and tablets to access information and search the Internet. Why? Because mobile is a medium where people want answers fast!

First, the form factor limitations of mobile phones don't accommodate a long list of links, and people wouldn't scroll through them anyway. Second, the huge growth in data and the failure of filters leave the door wide open for search services and apps that are "smart" enough to focus on a specific query and deliver genuinely useful results.

Tired of scanning the news every day to find relevant articles to read? Intelligent apps will do it for you. Services like this are based on machine learning, where the app tracks the user's click behavior and feedback to improve the precision of the service.

Another challenge is the increasing complexity of what people are searching for in the first place. Today's problems and questions involve many geographies, many disciplines, many communities, or, at the very least, many databases. From medical diagnoses to choosing the right car, people want the answers that suit them best. This is where AI comes in to enable approaches that infuse human preferences and judgments into search activity to deliver people advice, assistance, and what matters to them most.

To offer genuinely useful guidance to consumers making decisions, search apps must move beyond offering results based on popularity or page rank, and tap into many sources often, including social networks to deliver recommendations from trusted experts and communities. This approach also seeds a virtuous and lucrative cycle: Greater precision in the results (because they are based on real people with real opinions and influence, rather than databases) paves the way for equally high precision in mobile advertising and marketing, where the app is the channel to the customer and the trigger for an impulse buying decision.

FACT

A survey of 2,666 mobile users conducted by research firm TNS Global on behalf of Qualcomm found that if services were more personal and offered results more aligned with individual preferences and profiles, almost 60 percent of respondents said they would spend more time accessing content.

Similar to recommendations, services like virtual assistants will help tackle the information overload by identifying what is relevant and interesting to every single individual. To achieve this, apps will "learn" a customer's profile (drawing from a mix of implicit and explicit inputs) and alert them to content and information they would likely appreciate.

The advance of AI and microsensors paves the way for medical advice apps that provide personal and contextual diagnosis. By harnessing the knowledge of doctors with the use of AI, the result is personal, scalable, and far superior to services and apps that try to diagnose and help people based on single keyword and database matching. Feel a cold coming on? An interactive Q&A app on the mobile phone could allows users to input their symptoms—key variables such as their body temperature, the degree of nasal discharge (clear or colored), type of cough (dry or heavy)—and get back a diagnosis. A great opportunity for a trusted name in cough medicine or nighttime pain reliever to deliver people a branded app that they will use often.

ALERT

Apps are becoming more intelligent because they have to. Competition is heating up with AI-powered apps like Siri, Zite, and eBay's explorer for Windows 7 devices leading the way.

The First Step for Developers

AI is no longer rocket science. There are now tools and platforms that enable app developers to make intelligent apps without having to be experts in AI. This is where companies like Expertmaker are leading the way. Expertmaker provides a complete platform for intelligent solutions. In practice, the app developer only needs to plug the data and/or knowledge into the platform. The platform has a tool for modeling knowledge and data, a cloud server for processing, data mining capabilities, and an API for creating great user interfaces.

Say you want to create an app that is a more intelligent version of Flipboard, the popular and personalized social news magazine for iOS and Android. Your app would have to be able to scan articles and pull out all the ones that an individual user would likely be interested in. AI would come in to make the perfect match between content and user, bringing in intelligence in the form of text classification, feature extraction, optimization algorithms, machine learning, and other capabilities to "understand" the user, "know" what they want to read, and "pick" a perfect match based on this information.

AI is new to the mobile space and there is no limit to its applications. Use your imagination. How about a retail app that helps women choose the right cosmetics? They could input hair color, eye color, skin color, and personal style to help decide the combination that is right for them. Or a banking app that offers customers a virtual assistant experience with a virtual sales assistant who can alert them to products that fit with their current life situation (like looking for a house when a new baby arrives, for example). Think of scenarios where people need and value an informed opinion and you have a surefire idea for an app.

Where is the opportunity? Potentially everywhere. The obvious area of opportunity is retail and e-commerce. It's the Amazon model all over again. Only this time it's not enough to use collaborative filtering to tell the user that "people like you chose this item" and then proceed to recommend an item to buy. People demand to know what is truly right for them alone.

Interestingly, enterprises and companies that are consumer facing, such as department stores, fashion retailers, and luxury brands, are realizing that they can delight their customers by helping them make decisions. Put another way, these companies are in need of intelligent apps that are capable of providing personalized recommendations.

However, personalized recommendations are just the low-hanging fruit. There is also a requirement for apps that look at their existing data to uncover "hidden knowledge" that can be harnessed to better operate their business. This could be anything from improving the user experience of their online services to boosting process optimization.

ESSENTIAL

Enterprise companies sit on huge stockpiles of data about their products and services. This is where you can bring value by marrying the information these companies have stored in databases with intelligence in your app.

Moving forward, AI will likely do more than assist people to make decisions in their daily lives. It will power a new breed of precision search engines, combining parameterized search (searching according to specific parameters) and keyword search, allowing people to effectively have a conversation, rather than run a search, with the help of their intelligent apps.

As a result, searching will become finding as people interact with the search engine, and refining queries (enhancing precision) in response to helpful suggestions (parameters). High precision is possible because the technology is designed to predict optimal matches for each individual customer based on his or her personal needs.

Apps for Good

In just a few years, mobile devices will be exponentially smarter than they are today. More importantly, they will also be more contextually aware, which means they will "know" to present information and assistance to users at the moment they need it. This always-on/always-aware capability will grow to transform health care, paving the way for organizations, hospitals, physicians, and caregivers to deliver truly personalized care.

Already, analysts are predicting explosive growth in digital health apps. According to ABI Research, the market for mobile health apps is expected to quadruple to $400 million by 2016. To date, somewhere between 20,000 and 30,000 mobile health apps are available to users.

The vast majority are currently aimed at people who want to track their training or fitness routines, recording how far they run, how many calories they burn, or how much weight they lose. Another category of popular apps encourages social gaming to make staying healthy fun by allowing users to compete with family, friends, or online buddies on increasing exercise.

Apps can also enable home monitoring, where apps are connected to an organization allowing a form of remote patient monitoring for the elderly or disabled, for example. A recent IMS Health report forecast that 80 percent of the remote monitoring market will be mobile by 2016.

Enabling Caring and Sharing

At the other end of the spectrum, there is also a requirement for apps that simply keep people company, thus helping the healing process, easing the difficulty of living with a disease or disability, or just decreasing the feeling of social isolation among the elderly.

This is the course that Stuart Arnott, director of Mindings, a U.K. company focused on enabling "disruptive care" has purposely chosen. In technology, "disruption" is where a new and agile product or service comes into the market, displaces the incumbent, and increases the overall market.

Social care is an area where huge (often governmental) organizations run services.

Mindings is a service that enables people, from their mobile phone, to share captioned photos, text messages, calendar reminders, and social media content. In the case of Mindings, with just these four types of digital content, less is more—and all the more accessible for the audience of elderly and isolated for whom the app was designed in the first place.

Two functions stand out because they illustrate how small tweaks can make a huge difference. The captioning of photos is important because it gives the picture and the user context. If the user has a memory problem, and there are 820,000 people in the United Kingdom alone who are living with dementia, this contextual information is vital. Without putting a name to a face, the picture might be meaningless.

In addition, the "GotIt!" icon allows important feedback that enhances simple text messaging. In practice, when a family member sends a text or other content, the recipient need only touch the "GotIt!" icon on the screen to confirm they did indeed receive it. That confirmation is a welcome sign to the family member that the recipient is okay and interacting with the world around them.

The Mindings app is an easy blueprint other app developers can follow. Building a successful app to deliver social care is not about cool features and technology; it's about delivering what your user will appreciate.

It's also about addressing specific needs in a unique way. There are many opportunities where an app can improve quality of life.

Apps Transform Health Care

Clearly, mobile is impacting how people manage their own health and well-being. It's also having a huge effect on how medical professionals administer care to patients. Already, more than 80 percent of physicians own a mobile device, compared to 50 percent of the general U.S. population. About 30 percent of physicians are using smartphones and tablets to treat patients.

How will apps mobilize and personalize health care? Derek Newell, CEO of Jiff, a health care social network, has identified ways apps and smartphones will leave their mark, such as improving access to care. In the future, doctors' offices might be obsolete. Instead of having the government or insurance companies dictate that a visit must be in person, patients and physicians will decide together when a visit is best done live and when health care services can be delivered virtually. In addition, apps will be able to notify you when your doctor is running late, keep a tally of the number of pills you have taken and when you need to take them again, and refill your prescription for you.

Apps will also produce large call centers with house nurses, doctors, pharmacists, and other health care professionals who will watch, manage, and respond to this inbound data. In addition, digital health apps will allow providers to coordinate patient care in a complex environment. Say goodbye to Medicare fraud, as digital apps will allow Medicare to correlate claims data with location and time data from the digital health apps to look for fraud.

FACT

A ninety-nine-cent wheelchair-friendly application, dubbed TalkRocket Go, has been launched by MyVoice. The app, designed especially for people with speech disabilities, can help patients speak phrases and words by linking switches and buttons to mobile devices such as an iPad or iPhone.

In the future, everything that can be done digitally *will* be done digitally. Digital health apps will schedule appointments, tell you the doctor is running late, help monitor medications' side effects, and help you follow your care plan accurately. These changes will engage patients with their health and health care in new ways and radically reform health care delivery.

Sections of this chapers were contributed by Derek Newell, Jiff, Inc.; Martin Rugfelt, Expertmaker; Stuart Arnott, Spark Inspires.

APPENDIX A

Glossary of Mobile Terms

3G

Short for "third generation," the term is used to represent the third generation of mobile telecommunications technology.

A2P

Application-to-person.

AI

Artificial intelligence.

API

An application programming interface is a way for developers to access services from another company via computer programming.

App

A mobile app is a piece of software specifically designed to run on a mobile device, such as a smartphone or tablet.

Apple UDID

Unique device ID, an identifying number proper to each individual Apple consumer device that provides the ability to track the consumer across each app that is downloaded or app interaction.

AR

Augmented reality.

ASO

App store optimization.

Bespoke functionality app

A bespoke functionality app is an app that is designed to provide a solution to a specific need or problem.

Blog

Short for "web log," a blog is a web page that serves as a publicly accessible personal journal for an individual. Typically updated daily, blogs often reflect the personality of the author.

Cookie

A way for companies to track users' behavior on the Internet.

CPA

Cost per acquisition.

CPC

Cost per click.

CPD

Cost per unique download.

CPM

Cost per thousand.

CTR

Click-through rate.

Data-driven app

This is an app where the data is dynamic. This means it's either stored in a local (on the device) database or retrieves the data from an external source.

Device app

A device app is an app that makes use of the hardware to provide its core functionality. That means it taps into something based on a part of the device, such as the camera, accelerometer, or GPS.

Forum

An Internet message board for sharing ideas or opinions with others.

Google device ID

A unique ID that is associated with each individual Google consumer device and provides the ability to track the consumer across each app that is downloaded or app interaction.

GPS

Global positioning system, an accurate worldwide navigational and surveying facility based on the reception of signals from an array of orbiting satellites.

HTML

Hypertext Markup Language, a standardized system for tagging text files to achieve font, color, graphic, and hyperlink effects on World Wide Web pages.

IAP

In-app purchase.

JavaScript

A scripting programming language most commonly used to add interactive features to web pages.

LBS

Location-based services.

M2M

Machine-to-machine.

MMS

Multimedia messaging service.

Mobile web apps

Mobile web (or web app) experiences are quick and easy for small teams to implement and maintain, and as a result often only offer a slice of the full web content.

Native app/wrapper

Small application (small in data size) that is downloaded and installed on a mobile device. Content (such as pictures, etc.) is pulled over the Internet via a mobile data connection (e.g., Wi-Fi), and once the content is embedded in the service, the data connection can be closed and the content viewed offline. In essence, you are downloading a piece of content to your phone.

NDA

Nondisclosure agreement.

OS

Operating system.

Platform

A platform is an underlying computer system on which application programs can run.

Push notifications

Push notifications allow you to send messages directly to the people who have installed your app, even when the app is closed on a device.

Responsive design

Best suited to projects starting from scratch, these experiences require higher initial implementation costs, but allow for the broadest content scope and easiest maintenance.

ROI

Return on investment.

RSS feed

RSS is the acronym used to describe the de facto standard for the syndication of web content, including news feeds, events listings, news stories, headlines, project updates, and excerpts from discussion forums or even corporate information. A website that wants to allow other sites or apps to publish some of its content creates an RSS feed, allowing the distribution of that content. Syndicated content can include data.

SDK (software developers kit)

A set of tools, including API, frameworks, interface elements, and so on, used to create software, that is, apps.

SMS

Short message service.

SoLoMo

Social, location, and mobile.

SoundCloud

Social sound platform where anyone can create sounds and share them everywhere.

SS7 roaming network

On the public switched telephone network (PSTN), Signalling System 7 (SS7) is a system that puts the information required to set up and manage telephone calls in a separate network rather than within the same network that the telephone call is made on. Signalling information is in the form of digital packets. SS7 uses what is called out-of-band signalling, meaning that signalling (control) information travels on a separate, dedicated 56 or 64 Kbps channel, rather than within the same channel as the telephone call. Using SS7, telephone calls can be set up more efficiently and with greater security.

SSL/TLS

TLS (transport layer security) and SSL (secure sockets layer) are cryptographic protocols that allow making secure connections from a client to a server with SSL capabilities.

UDI

Unique device identifiers.

UI

User interface.

URL

Uniform resource locator.

UX

User experience, or how a person "feels" about using a product, service, or system.

Web service back-end

A system that maintains the data and/or business rules of a specific domain. Usually, it provides a specific role or has a specific responsibility in a system landscape.

Wi-Fi

Wireless local area network: a local area network that uses high-frequency radio signals to transmit and receive data over distances of a few hundred feet; uses ethernet protocol.

WPS

Wi-Fi-based positioning system.

Contributor Biographies

Akash Sureka, Hoopz

Akash Sureka is the founder and CEO of Hoopz Planet Info Pvt Ltd. A mobile industry veteran and leader, Akash has spent more than twelve years working on disruptive mobile technologies and applications. He's the inventor of new "Contextual Auto Search and Auto Web Discovery" technology and solutions for mobile and tablets.

Follow him on Twitter @akashsureka. Visit *www.hoopz.in*.

Alfred De Rose, Tego Interactive

Alfred began his career in mobile at a Prague-based development company that produced some of the biggest mobile game titles of the JavaME era. During this time he oversaw 100+ projects for companies like EA, Sony, Motorola, Gameloft, T-Mobile, and other game publishers and mobile network operators around the globe. In 2008, Alfred cofounded Tego Interactive with colleague Brian Avery to help clients understand and execute mobile strategies that meet concrete business needs.

Follow Alfred on Twitter @AlfredDeRose. Visit *www.tegointeractive.com*.

Andrew Bovingdon, Bango

Andy leads Bango's product strategy and is responsible for the interface between products and markets. He has over twenty-three years' experience in the global Internet market and has integrated a wide range of web technologies into the X-desktop product range for IXI Ltd. He has introduced the world to the web-based desktop (Webtop) at SCO; and designed the first web-based remote application access solutions for Tarantella Inc. Andy holds a bachelor's of science in computer science from Staffordshire University.

Follow Andy on Twitter @MrBov. Visit *www.bango.com*.

Asif Khan, LBMA

Asif is a veteran tech start-up, business-development, and marketing entrepreneur with nearly fifteen years' experience in the industry. He is currently working as a consultant, speaker, and venture capitalist. Asif recently formed the Location Based Marketing Association, an international group dedicated to research and education in location-specific marketing. Asif worked with companies such as Limited Brands, IBM, Baxter Pharmaceuticals, Molson-Coors, Communispace, Best Buy, American Airlines, Cineplex

Entertainment, ING Bank, and Sears. He holds degrees in economics and management sciences from the University of Waterloo.

He blogs at *www.betakit.com* and *www.streetfightmag.com*. Follow him on Twitter @AsifRKhan.

Chris Jones, CodeNgo

Chris Jones is the cofounder of CodeNgo, a start-up offering marketing and distribution services to help developers make more money with their apps. Chris has seventeen years' experience in marketing, working with brands including Adidas America, Boost Mobile, and Virgin Mobile. Chris is a graduate of Georgetown University and Northwestern University's Kellogg School of Management. He also formerly cochaired the Mobile Marketing Association's Urban Special Interest Group. He currently lives in Sydney, Australia, and is married with three children.

Follow Chris on Twitter @cjones2002 and @CODE_NGO. Visit *www.codengo.com*.

Dan Appelquist, BlueVia

Dan is the head of Product Management for BlueVia, Telefónica's developer portal and API platform. A London-based American expatriate, Dan has been a leading advocate of and participant in web standards through his work in W3C. He is a frequent speaker on technology and industry topics as well as a community and events organizer, having cofounded Mobile Monday London, the Mobile 2.0 conference series in San Francisco, and the Over the Air hack days in the United Kingdom.

Follow Dan on Twitter @Torgo and @BlueVia. Visit *https://bluevia.com/en/*.

Derek Newell, Jiff, Inc.

Derek is the CEO of Jiff, a health care technology company that is connecting the digital health ecosystem to the traditional health care system. He is also a managing director at HT3, a consulting firm dedicated to working with the purchasers of, developers of, and investors in new and emerging technologies. Prior to Jiff, Derek was the president of Robert Bosch Healthcare. Derek has degrees in molecular and cell biology, business administration, and public health, all from the University of California at Berkeley.

Follow Derek on Twitter @dereknewell. Visit *www.jiff.com*.

Ed Vause, Appromoter

Ed started his career in application software development at PlanIT Software in 1994 and launched his own software development business in 1998. In 2000 Ed cofounded PRshots.com, running the business from 2005 until 2011. PRshots distributes hi-res fashion and lifestyle photography to over 30,000 media users. In 2010, he was responsible for PRshots' launch in Germany, and cofounded Appromoter in 2011 with Jacki Vause, John Ozimek, and James Kaye to help app developers get their creations in front of the media.

Follow Ed on Twitter @Appromoter and @DimosoAgency. Visit *www .appromoter.com*.

Gary Schwartz

Gary is the CEO of Impact Mobile, Inc., North America chair of the Mobile Entertainment Forum (MEF), and author of *The Impulse Economy* and the upcoming book *Fast Shopper, Slow Store*. In 2006, Gary founded the mobile committee for the Interactive Advertising Bureau and has since worked to publish literature such as the *Mobile Buyer's Guide*, for which he received an IAB award for industry excellence in 2009.

You can read his book and blog at *www.theimpulseeconomy.com*. Follow Gary on Twitter at @impulseeconomy.

Heini Vesander, Blaast

Heini is responsible for communications and developer partnerships at Blaast. The company uses the revolutionary approach of running mobile apps in the cloud, and not on the device, and brings rich, always-on apps to feature phones and entry-level smartphones. Heini is from Helsinki, Finland, and an active member of the local start-up scene. She is part of the organizing team of Slush, the largest start-up event in the Nordics, Baltics, and Russia.

Follow Heini on Twitter at @Heinider or @Blaast.

Helen Keegan, Heroes of the Mobile Fringe

Helen has specialized in mobile for more than ten years and is a specialist consultant in mobile marketing, advertising, and media, working on a wide variety of projects. She also advises mobile start-ups on their business strategy and marketing efforts and helps mobile network operators and handset manufacturers with their developer relations programs. She has a regular column in *Mobile Marketing* magazine, and when she has time, she blogs about mobile marketing at *www.technokitten.com.*

Follow Helen on Twitter @technokitten.

Itay Godot, Inneractive

As an experienced publisher, Itay joined the Inneractive team in 2010 as VP of publisher relations. Now VP of marketing, Itay's responsibilities include forming and developing a dynamic and robust strategy to disrupt the exploding mobile monetization space. Before joining Inneractive, Itay cofounded Cellfer, a company that brought a Kindle-like reading experience to mobile handsets. In his free time, Itay is an avid mountain biker, and he likes traveling and downloading new and fun applications.

Follow him on Twitter @Itay_Gadot. Visit *www.inner-active.com.*

James Coops, MobyAffiliates

James has created a number of ringtone and mobile content services and built the early Indie mobile app store mjelly.com. He has consulted for mobile operators, media companies, agencies, start-ups, and investors in the United Kingdom and internationally. He has written for leading publications including *Techcrunch Europe*, *New Media Age*, *Corante.com*, and *Mobile Industry Review*. James now runs mobyaffiliates.com.

Follow James on Twitter @mobyaffiliates. Visit *www.Mobyaffiliates.com.*

Jennifer Hiley

Jennifer has over twelve years' digital experience, with eight years passionately specializing in mobile. She spent five years in Australia working as a mobile integration strategist for the mobile marketing agencies 5th Finger, TigerSpike, and the Hyperfactory. Jennifer has developed mobile strategies for blue-chip companies including Disney, Pernod, Coca-Cola, J&J, Land Rover, Lynx, and the Chartered Institute of Accountants in Australia. She then joined WeLoveMobile and recently launched Women in Wireless, where she serves as head of the Social Committee.

Follow Jennifer on Twitter @jenniferhiley.

Jez Harper, Tús Nua Designs

Jez Harper is the cofounder and CEO of Tús Nua Designs, an Irish app development firm. He has worked as a technical architect, project manager, and developer in the financial sector. He has provided software development consultancy services to businesses of all sizes, from entrepreneurs to global organizations. Jeremy also spends a great deal of time advising businesses on all aspects of their app requirements. Jeremy likes trees, mountains, and lakes. And Toblerones, he loves Toblerones. Follow Jez on Twitter @tasnuadesigns. Visit *www.tusnuadesigns.net.*

Jonathan Kohl

Jonathan has applied software testing, business analysis, and technical design and project management on mobile application projects for smartphones and tablets. He is an internationally recognized consultant based in Calgary, Alberta, Canada, where he is also the founder and principal software consultant of Kohl Concepts, Inc. He is the author of the book *Tap into Mobile Application Testing* (2012).

Follow Jonathan on Twitter @Jonathan_Kohl. Visit *www.kohl.ca.*

Joy Liuzzo, Wave Collapse

Joy Liuzzo is the president of Wave Collapse LLC. Combining data analysis, survey techniques, and proven approaches to increase sales, Joy continues to push her clients, and the industry, forward. Prior to launching Wave Collapse, Joy was vice president at InsightExpress. While there, she built the mobile research practice from the ground up. While in this role, Joy was elected as a North American board member to the Mobile Marketing Association.

Follow Joy on Twitter @joyliuzzo. Visit *www.wavecollapse.com.*

Ken Herron

Ken is a frequent author and speaker, and was the chief marketing officer for global communications service provider Purple Communications, social solutions and service provider SocialGrow, and the world's largest gay social network, MANHUNT. Prior to that, Ken was the vice president of interactive strategies for a global real estate franchisor Realogy. Ken earned his master's of international management from the Thunderbird School of Global

Management in Glendale, Arizona, and his bachelor's in international economics and German from Drew University's College of Liberal Arts in Madison, New Jersey. Living in Northern California, Ken is a contributing editor to mobile technology website MobileGroove.com, and the weekly cohost for *The LinkedIn Lady* radio show.

Follow Ken @kenherron.

Linda Daichendt, MTAM

Linda Daichendt is the executive director of the Mobile Technology Association of Michigan, a cofounder of Mobile Monday Michigan, and a cohost/coproducer of the international mobile marketing podcast series, "The Mobile Marketing Review." She is a marketer and blogger, and one of Michigan's leading proponents of the use of mobile technologies for businesses of all sizes. Linda's daily focus is on educating Michigan business, government, and education stakeholders about the opportunities that mobile technology provides.

Follow Linda on Twitter @GoMobileMI. Visit *www.gomobilemichigan.org*.

Lisa Ciangiulli, Optism

Lisa Ciangiulli is director of marketing at Alcatel-Lucent and is responsible for leading the marketing strategy for the company's Mobile Commerce solutions. Lisa has held positions in both marketing and strategy with worldwide responsibilities. In her current role, Lisa is leading the global marketing strategy for the company's Mobile Commerce portfolio. Lisa holds a BA in business and economics from the University of Pittsburgh.

Follow Lisa on Twitter @LisaCiangiulli. Visit *www.optism.com*.

Magnus Jern, Golden Gekko

Magnus is CEO and founder of Golden Gekko, a mobile app developer. He has over ten years' experience with content strategies, online marketing, search, location-based services, app development, and mobile marketing for global consumer brands. Prior to founding Golden Gekko, Magnus worked as a senior manager at Vodafone. Before that Magnus worked in several marketing and strategy roles at Vodafone, Orange, Driftbolaget, and Framfab. Magnus is a frequent speaker at conferences and events about the future of mobile apps and service. Follow Magnus @MagicMagnus. Visit *www.goldengekko.com*.

Martin Rugfelt, Expertmaker

Martin is a mobile entrepreneur with mobile commerce, instant shopping, app development, and mobile payments. Martin also worked at Orange Group as director of product solutions. Martin is now chief marketing officer at Expertmaker, an artificial intelligence software platform company with a vision of making the Internet more intelligent.

Follow Martin on Twitter @expertmakertool. Visit *www.expertmaker.com*.

Martin Wilson, Mobileweb Company

Martin is the managing director and cofounder of Mobileweb Company. He has a background in marketing, business development, and strategy, and has been involved in digital media for over fifteen years. Prior to Mobileweb Company, Martin headed the independent mobile consultancy, Indigo102. He spent almost twelve years at Yell, one of the world's leading directory publishers. He holds a bachelor's degree in economics and business studies from the University of Wolverhampton.

Follow Martin on Twitter @indigo102 and @mobilewebco. Visit *www .mobileweb.co.uk*.

Matos Kapetanakis, Vision Mobile

Matos is the marketing manager of VisionMobile, a market strategy analysis firm. His role involves overseeing the marketing, communications, and PR efforts of the company. Matos is also the project manager of the Developer Economics research series, as well as other developer research projects. He is a physics and MBA graduate and has experience as a marketing manager in cultural media companies.

Follow Matos on Twitter @visionmobile. Visit *www.visionmobile.com*.

Matt Lutz, AppClover

As cofounder and COO of AppClover.com, Matthew, along with his business partner and CEO, Len Wright, are working at creating and growing their online Global App Community, along with publishing *Appreneur Magazine*, the world's first monthly mobile magazine dedicated to app marketing and monetization, all the while launching their online platform—Appzine Machine—which allows people to easily publish their own mobile magazine app via Apple's Newsstand. Matthew graduated from the College for Creative Studies in Detroit.

Follow Matt on Twitter @AppClover. Visit *www.appclover.com*.

Mike Anderson, Chelsea App Factory

Matt is an industry veteran who has successfully developed businesses, created new brands, and managed start-ups. Matt has held management positions at the *Sun*, the *News of the World*, the *Evening Standard*, and *Metro*. He started the Chelsea Apps Factory two years ago. Today Chelsea Apps Factory specializes in developing business enterprise apps. Its clients include Deloitte, GSK, Waitrose, BP, Mumset, and Vodafone.

Follow Mike on Twitter @ChelseaApps. Visit *www.chelsea-apps.com*.

Moshe Vaknin, YouAPPi

After beginning his career at AT&T Bell Labs, Moshe founded three successful start-ups before starting YouAPPi, a cross-platform app recommendation and distribution solution. Moshe has over twenty years of experience developing and marketing solutions in the fields of mobile communication, advertising, and marketing, holding VP and C-level positions at leading organizations in these industries.

Follow Moshe @YouAppi. Visit *www.youappi.com*.

Paolo De Santis, Chupamobile

Born in Rome, Paolo has worked with companies including Nike, Nokia, Telecom Italia, Enel, Twentieth Century Fox, and Ducati. He cofounded Dlite, a digital marketing agency that has worked with international brands, with offices in Rome and in Dubai, and Chupamobile in 2011, with the mission of connecting mobile developers, providing them a new way to generate revenue and speed up apps' development time. He loves technology, football, lasagna, and surfing in his spare time!

Follow Paolo on Twitter @Chupamobile. Visit *www.chupamobile.com*.

Paul Poutanen, Mob4Hire

Paul is a Calgary-based high-tech executive and entered the mobile technology field after a long stint as a senior management consultant with Ernst and Young. He has worked in management for wireless hardware and cellular location firms such as Wi-LAN and Cell-Loc. Paul became president of Blister Entertainment, where he developed the first mobile location-based games for North America. Paul launched Mob4Hire in 2007. He is an industrial and mechanical engineer.

Follow Paul on Twitter @mob4hire. Visit *www.mob4hire.com*.

Phil Hendrix, immr

Dr. Phil Hendrix is the founder and director of immr (*www.immr.org*), a research and advisory firm, and an analyst with GigaOm Pro. As an analyst, Phil focuses on mobile innovation and the implications for companies across industries. Before founding immr, Phil was a partner with Diamond-Cluster (a strategy and technology consultancy, now part of PwC), founder and head of IMS (Integrated Measurement Systems), and a principal with Mercer Management Consulting (now Oliver Wyman). He has held faculty positions at Emory University and the University of Michigan.

Follow Phil on Twitter @phil_hendrix. Visit *www.immr.org*.

Rimma Perelmuter, MEF

As MEF (global community for mobile content and commerce) executive director, Rimma leads the organization in its mission to shape the industry, connect thought leaders, drive monetization opportunities, and provide competitive advantage to members. MEF is a member network with international reach and strong local representation.

Follow Rimma on Twitter @rimma1. Visit *www.mefmobile.org*.

Rob Woodbridge, Untether TV

Rob has had roles ranging from strategic advisor, board member, and coach to VP of operations, president, and CEO. Rob has helped shape strategy, marketing initiatives, and product development to extend existing business into the mobile world. Rob's experience includes a mobile game company targeting consumers, a mobile IT solution targeting enterprises, and a video podcast focused on helping entrepreneurs build new empires or companies extend revenue through mobile and pervasive computing.

Follow Rob on Twitter @robwoodbridge. Visit *www.untether.tv*.

Ryan Morel, PlacePlay

Ryan is the CEO of PlacePlay, which provides iOS and Android app developers a solution to increase app revenue and drive installs. Ryan has been in the mobile gaming industry since 2005. Ryan lives in Seattle, where PlacePlay is headquartered.

Follow Ryan on Twitter @ryanmorel. Visit *www.placeplay.com*.

Sam Chan, WIP

The WIPsters (Wireless Industry Partnership) are based in Vancouver, London, and Austin, and use their knowledge, charm, and good looks to connect mobile developers to people, information, and resources to increase innovation and market success. Find Sam and other WIPsters at their events such as WIPJams, Muthers!, and DroidconUNITED KINGDOM; and check out their online resources including the Developer Marketplace, AppStore and API Catalogs, and Global Mobile Community Calendar.

Follow Sam on Twitter @anothersamchan and @wipjam. Visit *www .wipconnector.com*.

Scott Townsend, Urban Airship

Scott Townsend is the director of Marketing at Urban Airship, a leader in push messaging. Before joining Urban Airship, Scott was a marketing manager at Google, where he helped build digital strategies and educate companies on how to empower their brands and drive business goals using new media. He has worked with large global brands including Burger King, Campbell's, Coca-Cola, Mars, Papa John's, Unilever, and many others.

Follow Urban Airship on Twitter @urbanairship. For more information, visit *www.urbanairship.com*.

Stuart Arnott, Spark Inspires

Stuart is a multimedia producer with seventeen years of experience. After the birth of his daughter, Stuart developed Mindings, which enables the sending of personal caption photos, text messages, calendar reminders, and social media content from a mobile phone to a digital screen that the user doesn't even need to touch. Mindings has already been recognized by Cambridge University with a Cambridge Wireless Discovering Startups 2011

Award and was recently cited by TechRadar as one of the top twenty UK start-ups to look out for.

Follow Stuart on Twitter @MindingsStu. Visit *www.sparkinspires.com.*

Suzie Mitchell

Suzie Mitchell is founder and CEO of Clear Writing Solutions, which helps health IT companies market to boomers, seniors, and caregivers. She is the author of the BoomerTech blog and of a weekly blog for AARP called App of the Week. She is a blog contributor for *UX Magazine*, *MobileGroove*, *Rock Health*, *Health 2.0*, and *Aging 2.0*. She is president of Mitchell Research and Communications, Inc., a marketing research, public relations, and public affairs firm. She is also coauthor of the book *Growing into Grace: Adventures in Self Discovery through Writing*.

Follow Suzie on Twitter @suziemitchell. Visit *www.mitchellpr.com.*

Viki Zabala, Fiksu

Viki serves as the director of marketing at Fiksu, Inc., developer of Fiksu, a mobile apps marketing platform that uses advanced media optimization technology. Prior to joining Fiksu, Viki held marketing leadership roles across a variety of industries, including computer software, video, online media, advertising, and mobile marketing. Viki was recently recognized for her contributions and continued influence in the mobile marketing community as one of *Mobile Marketer*'s 2012 Mobile Women to Watch.

Follow Viki on Twitter @vikipierce and @fiksu. Visit *www.fiksu.com.*

Yasmina Haryono, Fjord

Yasmina creatively manages design teams and steers projects from vision to deployment phases. She enjoys helping clients discover their inner Venn, designer, storyteller, and/or model-thinker. She holds degrees in interaction design and has experience designing strategies, products, and services for clients including Philips Design, Vodafone, Experientia, Nokia, and Turkcell.

Follow Yasmina on Twitter @yasmina. Visit *www.fjordnet.com.*

INDEX

Abecassis, Arie, 127
Ad funding, 170
Advertising. *See also*
  Marketing
    ad exchanges, 180–81
    ad networks, 170, 175–76,
      180–86, 214
    banner advertising, 175
    challenges for, 181–84
    in-app advertising, 175–76
    revenue from, 181–85
    sponsors for, 179
    tips for, 181–88
    video advertising, 177
Age Wave, 51
Anderson, Mike, 67, 160, 292
Android OS (Google), 20–22,
  62, 64–65
Ansari, Yasser, 251
App charts, 38–39, 153
App discovery, 132, 142–43,
  189–204
App Economy, 12, 61–62
App review sites, 60, 200–201
App stickiness, 166–67
App stores
    approval for, 151–52
    choosing, 146
    for enterprise apps, 158–59
    healthy app stores, 156–57
    major stores, 148–51
    niche stores, 147
    operator stores, 147–48
    payment via, 157–58
    policies of, 165
    priorities for, 159–60

ratings charts for, 152–54
  shortcomings of, 155–56
  submitting to, 149–53
Appelquist, Dan, 286
Apple iOS, 20–21, 30–31, 62, 64
Application programming
  interfaces (APIs), 22, 149, 191,
  249, 260, 272
"Appreneurs," 250–59
Apps
  basics of, 14–15
  building, 55–67, 69–82
  core areas of, 26–28
  crashing of, 240–48
  customers for, 22–26
  definition of, 14–15
  designing, 95–106
  development of, 26–27,
    83–94
  distribution of, 26, 28
  downloading, 33, 37
  earnings from, 63–66, 167–
    68
  environment of, 26–28
  explanation of, 14–20
  future of, 263–75
  ideas for, 22–26, 37–39
  maintaining, 237–48
  number available, 30
  offering, 24–26
  role of, 22–26
  testing, 119–28
  types of, 15–18, 31–33, 70
Arnott, Stuart, 274, 276, 294–95
Artificial intelligence (AI),
  269–74

Audience
  by age, 42–46
  targeting, 22–26
  understanding, 22–26, 41–53

Baby Boomers, 42, 48–52
Bacon, Jeff, 194
Bada OS (Samsung), 20
Banner advertising, 175
Bespoke functionality app, 70
BlackBerry OS (RIM), 12, 20–
  21, 30, 63–64
Bloggers, 47, 60, 197–98, 200–
  201
Bovingdon, Andrew, 160, 172,
  285
Brin, Sergey, 46
Building apps. *See also*
  Designing apps; Developing
  apps
    budget for, 73, 84–85
    costs of, 69–73
    knowledge for, 69–82
    resources for, 56–57
    skills for, 85–86, 91–94
    strategies for, 55–67
Business models, 161–62,
  169–71, 229. *See also*
  Monetization models

Career, starting, 96
Case studies, 251–62
Chan, Sam, 67, 294
Ciangiulli, Lisa, 53, 290

Click-through rates, 113, 182

Client lists, 190

Communication skills, 75–76, 94

Community, building, 220–24

Compensation, 78–79

Competitors, researching, 37–39

Connectivity, 35, 231, 264

Constantly Connected group, 42–43

Consumer, protecting, 108–10. *See also* Privacy concerns

Consumer trends, 35–36

Content, updating, 19, 238–39

Contracts, 74–75, 80

Cooper, James, 288

Cooper, Martin, 13

Copyright considerations, 79–80

Cost-per-click (CPC), 175, 180

Costs of apps, 69–73, 89–90

Coupons, 47, 139, 176–77, 208–9

Crashing of apps, 240–48

Cross-platform apps, 18, 63, 66, 146, 180–81, 247–48

Crowd-sourcing, 45, 125–28, 247

Customer service, enhancing, 209, 266–68

Customers
database of, 190
first impressions for, 191–92
ideas for, 22–26

loyal customers, 190–91, 205–17
targeting, 22–26
understanding, 22–26

Daichendt, Linda, 67, 290

Data access, 20, 34–35

Data-driven app, 70, 72, 211, 241, 243

De Rose, Alfred, 94, 285

De Santis, Paolo, 292

Decision-making strategies, 35–36

Demographics, 25, 42–52, 266

Designing apps. *See also* Building apps; Developing apps
consistency in, 97–98
context of apps, 98–99
"do's and don'ts" for, 97
for global capabilities, 101–3
for multiple devices, 100–101
platform designs, 100–101
similarities in, 98
simplicity of design, 96–97, 105
speed of apps, 99–100
tablet platforms, 100–101
usage of apps, 98–99
user experience (UX), 105–6
user interface (UI), 96–105
visualizing apps, 104–5

Developer Economics 2012, 61–64, 146–47, 155, 204, 206

Developers
communication skills of, 75–76, 94
earnings for, 63–66
finding, 91–93
freelance developers, 87–88
in-house developers, 84–86, 88
meeting, 58–60
obstacles for, 66–67
skills of, 85–86, 91–94

Developing apps. *See also* Building apps; Designing apps
budget for, 73, 84–85, 90
core areas for, 26–28
costs of, 69–73, 89–90
in-house development, 27, 84–86
methods of, 86–89
outsourcing, 84, 87
skills for, 85–86
third-party developer, 27
time for, 86

Development agencies, 88–90

Device app, 70, 72

Digital Moms, 46–48

Digital Natives, 42, 44–45

Distribution of apps, 26, 28, 86, 146–47

Do It Yourself (DIY) tools, 87

Doerr, John, 135

Download metrics, 214–15

Download numbers, 30–31, 37

Dychtwald, Ken, 51

Earnings from apps, 63–66, 167–68

Engel, Joel S., 13

Exclusivity, 81

First impressions, 191–92

Freelance developers, 87–88

Freemium model, 163–64

Functionality app, 70

Future of apps
  artificial intelligence, 269–74
  customer service, 266–68
  health care, 274–76
  information overload, 270–72
  machine-to-machine market, 264–65
  personal recommendations, 272–74
  virtual assistants, 270–73
  voice recognition, 267–68

Game apps
  business models for, 169–72
  location services for, 143
  paying for, 169–71
  popularity of, 37, 43

strategies for, 198–99
  types of, 70

Gates, Bill, 46

Generation M, 42, 45

Gen-X, 42, 46

Gen-Y, 42, 46

Glossary of terms, 277–83

Godot, Itay, 188, 288

Google Android OS, 20–22, 62, 64–65

Google Play apps, 30

GPS feature, 20, 36, 136–38, 140–42, 244

Hackathons, 59

Hagan, Michael, 254

Harper, Jez, 82, 94, 106, 160, 248, 289

Haryono, Yasmina, 106, 295

Health care, 274–76

Hendrix, Phil, 143, 293

Herron, Ken, 235, 289–90

Hiley, Jennifer, 28, 106, 288

HP webOS, 20

HTML5 apps, 17, 167, 247–48

Hybrid apps, 18, 247–48

Ideas for apps, 22–26, 37–39

In-app advertising, 175–76

In-app payments, 165–66

Information overload, 270–72

In-house developers, 27, 84–86, 88

Intellectual property (IP), 79–80

Investors, 228–29

iOS (Apple), 20–21, 30–31, 62, 64

iPad, 12, 30–31, 36

iPhone, 20–23, 27, 30, 64

Ito, Mizuko, 53

Jern, Magnus, 82, 290

Jobs, Steve, 20, 46

Jones, Chris, 160, 286

Kapetanakis, Matos, 67, 291

Kaye, James, 287

Keegan, Helen, 235, 287–88

Kemski, Mike, 190

Khan, Asif, 143, 188, 285–86

Kohl, Jonathan, 128, 289

Legal considerations, 74–75, 79–82

Linux MeeGo, 20

Liuzzo, Joy, 143, 289

Location-aware marketing, 177

Location-based marketing, 138–43, 177–79

Location-based service, 36, 135–43

Loyalty, achieving, 190–91, 205–17

Lutz, Matt, 40, 188, 204, 291–92

Machine-to-machine (M2M) market, 264–65
MacLean, Bretton, 260
Maintaining apps
  challenges of, 238–39
  contract for, 239–40
  crash issues, 240–48
  crash reporting, 242–43
  healthy apps, 248
Market research
  by age, 42–46
  on Baby Boomers, 48–52
  of competition, 37–39
  on Constantly Connected group, 43–44
  demographics for, 42–52
  on Digital Natives, 44–45
  by generation, 42–52
  on Generation M, 45
  on Gen-X, 46
  on Gen-Y, 46
  on Millennials, 46
  revenue projection for, 39–40
  target audience, 41–53
  tools for, 38–39
Market validation, 38–39
Marketing
  advertising and, 173–88
  app discovery, 132, 142–43, 189–204
  five P's of, 41
  location-aware marketing, 177
  location-based marketing, 138–43, 177–79

loyal customers and, 205–17
  mobile marketing, 173–74
  networking, 219–35
  social networking and, 192, 219–35
  sponsors for, 179
  strategies for, 216–17
McCarthy, John, 269
MeeGo (Linux), 20
Meeker, Mary, 20, 36
Message services, 61, 136–37, 174–75, 206–9
Microsoft Windows Phone OS, 20–22, 30, 63–64
Millennials, 42, 46
Mitchell, Suzie, 53, 295
Mobile advertising, 173–74, 177. *See also* Advertising
Mobile app development agencies, 88–90
Mobile demographics, 42–52
Mobile marketing, 173–74. *See also* Marketing
Mobile phones. *See also* Smartphones
  first call from, 13
  growth of, 11, 20–21
  number of users of, 11, 20–21, 30, 37
  role of, 22–26
Mobile strategy, building, 23–24, 229
Mobile technology, 11, 13–14, 29
Mobile video, 177, 261
Monetization models

billing solutions, 162, 167
  credit card processing, 162, 167–68
  explanation of, 162–66
  freemium model, 163–64
  for game apps, 169–72
  in-app payments, 165–66
  pay-per-download, 162–63
  popular models, 168
  pricing content, 168–69
  repeat business, 166–67
  subscriptions, 164–65
Money matters, 63–66, 161–72
Morel, Ryan, 172, 183, 188, 294
Multimedia messaging service (MMS), 61, 136, 177

Native apps, 17, 167, 248
Networking, 58–60. *See also* Social networking
Newell, Derek, 275, 276, 286
Nokia Symbian, 21–22, 30, 63
Nondisclosure agreement (NDA), 74–75

Offer walls, 169–70, 178, 183–84
Offshore development, 87
Open API, 249, 260
Operating system (OS), 20–21, 82
Owens, Eric, 190
Ownership of code, 79–82
Ozimek, John, 287

Paananen, Vesa-Matti "Vesku," 11

Partners/partnerships, 58, 61–62, 66–67, 78, 229, 253

Pavlidis, Mark, 260

Payment models. *See also* Monetization models
freemium model, 163–64
in-app payments, 165–66
pay-per-download, 162–63
subscriptions, 164–65

Payment terms, 77–79

Pease, Bob, 120

Perelmuter, Rimma, 117, 293

Personal, Portable, Pedestrian, 53

Personal recommendations, 272–74

Pierce, Taylor, 39

Platforms
choosing, 63
cross-platform apps, 18, 63, 66, 146, 180–81, 247–48
designs for, 96–97, 100–101, 105
popular platforms, 62–63
understanding, 20–21

Postlaunch challenges, 238–39

Poutanen, Paul, 128, 292–93

Prebuilt apps, 93

Press coverage, 60

Press releases, 201

Priebatsch, Seth, 254

Privacy concerns
balancing privacy, 115–16
consumer protection, 107–11
data security, 114–15
privacy policy, 113–14
transparency and, 109–16
trust and, 111–13, 116

Privacy policies, 113–14

Promotion, 191, 197–201. *See also* Marketing

Push notifications, 174, 176, 207, 209–13

Quotes, giving, 74–77

Repeat business, 166–67

Resources, 27

RIM BlackBerry OS, 12, 20–21, 30, 63–64

Ringtones, 11, 146, 165

Roaming network, 35

Rugfelt, Martin, 276, 291

Samsung Bada OS, 20

Schwartz, Gary, 287

Security concerns
balancing privacy, 115–16
consumer protection, 107–11
data security, 114–15
privacy policy, 113–14
transparency and, 109–16
trust and, 111–13, 116

Sharma, Chetan, 264

Shopping apps
consumer behavior, 132–35
customer loyalty, 139–40
entertainment industry, 140–41
evolution of, 130
location data for, 135–43
monetizing apps, 141–43
music industry, 140–41
statistics for, 131
tracking shoppers, 138, 139
understanding shoppers, 130–31

Short Messaging Service (SMS), 61, 136–37, 174–75, 206–9

Sizing Up the Global Mobile Apps Market, 264

Smartphones. *See also* Mobile phones
first smartphone, 20
growth of, 11, 20–21
number of users of, 11, 20–21, 30, 37
operating systems of, 20–21
ownership by age, 42
platform for, 20–21
role of, 22–26

Social media, 34, 44–48, 224–28

Social networking
follow-up for, 232–35
goals for, 228–30
marketing and, 192, 226–27

online communities,
220–24
skills for, 227–28
understanding, 224–27
Software bugs, 92
Software development kits
(SDKs), 27, 91, 149, 180–81
Software tools, 92–93
SoLoMo, 135–36, 140
Source code, owning, 79–82
Sponsors, 179
Stores for apps, 145–60
Subscriptions, 20, 164–65, 209
Success
of "appreneurs," 250–59
case studies of, 251–62
tips for, 252–62
Sureka, Akash, 204, 285
Symbian (Nokia), 21–22, 63

Target audience, 22–26,
41–53
Technology
decisions based on, 35–36
evolution of, 11, 13–14
growth of, 29
Testing apps
crowdsourcing, 125–28
deletable offenses, 120–21
emulators for, 120
feedback for, 125–28
for local audiences, 127–28
perspectives on, 122–24
potential problems, 124–25

on various platforms,
121–22
Text messaging, 136–37,
206–9
Third-party resources, 27
Time management shift, 36
Townsend, Scott, 217, 294
Trends, 20, 30, 35–36, 253, 256

User experience (UX), 84–86,
105–6
User interface (UI), 71, 84–86,
96–105, 246–47

Vaknin, Moshe, 204, 292
Vause, Ed, 204, 287
Vause, Jacki, 287
Vendor-supported apps, 93
Vesander, Heini, 160, 287
Video advertising, 177
Virtual assistants, 270–73
Visualizing apps, 104–5
Voice recognition, 267–68

Web apps, 16, 63, 85
webOS (HP), 20
Wilson, Martin, 28, 291
Windows Phone OS
(Microsoft), 20–22, 30, 63–64
Woodbridge, Rob, 262, 293

Zabala, Viki, 217, 295

We Have
EVERYTHING®
on Anything!

With more than 19 million copies sold, **the Everything® series** has become one of America's favorite resources for solving problems, learning new skills, and organizing lives. Our brand is not only recognizable—it's also welcomed.

The series is a hand-in-hand partner for people who are ready to tackle new subjects—like you!

For more information on the Everything® series, please visit *www.adamsmedia.com*

The Everything® list spans a wide range of subjects, with more than 500 titles covering 25 different categories:

Business	History	Reference
Careers	Home Improvement	Religion
Children's Storybooks	Everything Kids	Self-Help
Computers	Languages	Sports & Fitness
Cooking	Music	Travel
Crafts and Hobbies	New Age	Wedding
Education/Schools	Parenting	Writing
Games and Puzzles	Personal Finance	
Health	Pets	